Buying and Selling the Environment

Buying and Selling the Environment

How to Design and Implement a PES Scheme

Gabriela Scheufele

Visiting Research Fellow
Crawford School of Public Policy
ANU College of Asia and the Pacific
Acton, Australia

Jeff Bennett

Emeritus Professor
Crawford School of Public Policy
ANU College of Asia and the Pacific
Acton, Australia

ELSEVIER

ACADEMIC PRESS
An imprint of Elsevier

Academic Press is an imprint of Elsevier
125 London Wall, London EC2Y 5AS, United Kingdom
525 B Street, Suite 1650, San Diego, CA 92101, United States
50 Hampshire Street, 5th Floor, Cambridge, MA 02139, United States
The Boulevard, Langford Lane, Kidlington, Oxford OX5 1GB, United Kingdom

BUYING AND SELLING THE ENVIRONMENT

Notices
Practitioners and researchers must always rely on their own experience and knowledge in
evaluating and using any information, methods, compounds or experiments described
herein. Because of rapid advances in the medical sciences, in particular, independent
verification of diagnoses and drug dosages should be made. To the fullest extent of the
law, no responsibility is assumed by Elsevier, authors, editors or contributors for any injury
and/or damage to persons or property as a matter of products liability, negligence or
otherwise, or from any use or operation of any methods, products, instructions, or ideas
contained in the material herein.

ISBN: 978-0-12-816696-3

Publisher: Candice Janco
Acquisition Editor: Graham Nisbet
Editorial Project Manager: Ruby Smith
Production Project Manager: Kiruthika Govindaraju
Cover Designer: Alan Studholme

Working together
to grow libraries in
developing countries

www.elsevier.com • www.bookaid.org

Getting started

1.1 What are we doing here?

Concerns among the general public, internationally, regarding the condition of the environment have grown over the past half century. The conversion of areas of natural ecosystem for agricultural, forestry, mining, industrial, and urban development together with the encroachment of invasive feral species and the increased prevalence of wildlife poaching has caused losses of biodiversity and even species extinction. An increasingly well-educated and wealthy public, fueled by the well-publicized campaigns of "green" activist nongovernment organizations have pressured governments to act to protect environmental assets by direct action such as the setting aside of conservation reserves.

However, government actions have not always been sufficient to appease community (and activist) concerns. Nor have government actions always been successful. Costs of action have blown out and outcomes have been disappointing. This has been particularly apparent in less developed countries, often the location of areas that are under the greatest pressure from development and where biodiversity assets are most bountiful. This is even more the case when those looking for environmental protection are people living in more affluent developed countries, outside the political jurisdiction of their concerns.

Because of this inadequacy of direct government intervention, frustrated interest groups and governments seeking improvements in their environmental management performance began searching for alternative ways of achieving their environmental protection goals. For a range of reasons, Payments for Environmental Services (PES) schemes have become one of the alternatives to rise in prominence.

This book is aimed at meeting this growing interest in PES schemes in two ways. First, it is designed as an instructional manual for practitioners, policy makers and their advisors in government and nongovernment organizations. It provides a step-by-step demonstration of both the design and build phases of PES schemes based on first-hand experiences gained through an implementation in Lao People's Democratic Republic (PDR). Hence, both the conceptual and practical hurdles that must be overcome in the creation of an operational and efficient PES scheme are detailed and addressed.

Buying and Selling the Environment. https://doi.org/10.1016/B978-0-12-816696-3.00001-6

Second, the book presents original research in applied economics and bio-economic modeling that underpins the design and construction of PES schemes. As such, the book seeks to alert practitioners and policy makers to the pitfalls likely to be encountered by PES schemes. These pitfalls have provided significant barriers to past attempts to implement PES schemes and a goal of this book is to clear the way for those seeking to establish future PES schemes. However, the book is not just focused on the problems to be faced in implementing a PES scheme. It is fundamentally aimed at demonstrating the promise of the approach.

1.2 How to use this book

Readers of this book are likely to come from a range of backgrounds. Environmental management professionals often have training and skills in disciplines ranging from ecology to political science. They also come equipped with a diversity of experience in field operations. This book is designed to accommodate that heterogeneity.

The intellectual core for the book lies in applied economics. That means some readers will not be adequately versed in some of the core principles that are used. To ensure those noneconomist readers are brought sufficiently up to speed to cope with the economic content, special stand-alone "concept boxes" are placed at strategic points in the book. This also enables readers with economics training to skip those parts of the book that will already be familiar to them.

The book is also structured around chapters that put together a step-by-step approach to designing and building a PES scheme. Necessarily, some of the steps are not entirely sequential but the chapters indicate exactly how the steps interrelate and set out the logistics required for implementation.

Most chapters start with a description of how the material fits into the step-by-step approach and then provides an outline of its conceptual base. Where necessary, this material is presented as a concept box. Each chapter proceeds to set out the relevant material in a generalized form that can be related to the widest range of applications. To show how the concepts and processes of application can be used in a specific context, each chapter uses the case study of a biodiversity protection PES scheme implemented in Lao PDR. Relevant material produced in the Lao PES scheme project is referenced in the respective sections and provided in the respective annex.

Key challenges, lessons learnt, and future possibilities are summarized at the end of each chapter. The aim is to give readers some "heads-up" warnings for particularly problematic stumbling blocks that are likely to be encountered on the way to implementing a PES scheme. However, it is also intended that readers can leave each chapter—and eventually the whole book—with some inspiration in the knowledge that by following the step-by-step guide provided a conceptually sound yet practical PES scheme is possible.

Each chapter ends with a section devoted to a list of "References and further readings." The lists will assist readers who are interested in taking their knowledge on specific elements of the material covered in each chapter to a higher level. This section also acknowledges the inputs and publications produced by team members and collaborators of the Lao PES project as well as authors external to the project, which formed the basis for each chapter.

The remainder of this chapter is devoted to providing a brief rundown on the structure of the book and setting out some key principles that underpin the approach taken in this book to designing and building a PES scheme. The goal is to provide the reader with a road map of where we are about to travel and a solid foundation on which to start the journey.

1.3 A road map

As this book's title implies, PES schemes are essentially about the introduction of market forces to the process of protecting the environment. PES schemes seek to create circumstances in which buyers and sellers can interact by paying for and being paid for environmental protection. The book is therefore structured around the key elements that constitute a market. The steps set out to design and build a PES scheme are thus classified within each of these market elements.

In Chapter 2, the context in which the PES "market" is to be established is detailed. This is not only a matter of defining the environmental service to be bought and sold but also embodies dimensions including space, time and the people or organizations likely to be involved as either buyers, sellers, or brokers (otherwise known as agents or intermediaries).

Chapter 3 shows how the biophysical backdrop of any PES scheme needs to be well understood before any overlaying of a "market" for environmental services. In particular, the relationship between the environmental management actions that potential sellers can perform and the environmental service outcomes that potential buyers are interested in must be considered. This is the same process that say a wheat farmer has to understand before becoming a grain seller: What amount and quality of wheat can be produced, given inputs of land, fertilizers, tractors, etc.? It is, essentially, a productivity analysis.

Chapter 4 looks at the buyer side of the market. Specifically, the extent of potential buyers' "willingness to pay" for the environmental service must be estimated to understand the strength of demand in the market. The techniques used to quantify willingness to pay are outlined, and the processes used to aggregate different sources of demand are given attention. Particular attention is given to one demand estimation technique known as Choice Modeling.

The seller side of the market is considered in Chapter 5. The extent of potential sellers' "willingness to accept" payments for them to produce the environmental service must be estimated to understand what supply may be available in a PES market and the nature and extent of incentives faced by potential sellers. The particular technique that is advanced for this stage of the PES process is known as a "conservation auction." Aggregation of willingness to accept estimates across multiple potential sellers is also a focus of attention for this chapter. A feature of the chapter is its treatment of the process needed to ensure that potential sellers and their broader communities are engaged in the PES scheme process. This is especially critical in developing country contexts where those best able to act as sellers may be only poorly educated and have limited exposure to conventional markets, let alone the mechanics of a PES scheme. Furthermore, such potential sellers may have to learn to carry out environmental management tasks of which they have little or no experience.

With buyer and seller behavior set out in the two preceding chapters, Chapter 6 is devoted to bringing the two sides of the PES market together. This is a key chapter as it leads to the setting of a "market price" for the PES scheme. This is the amount paid by buyers and received by sellers. There are significant complexities in bringing together the information collected on buyers' willingness to pay and sellers' willingness to accept because of the different units of measurement used for the different market components. Buyers pay for environmental services but sellers are usually paid for the actions they take to supply them. A key part of this chapter is therefore concerned with gaining an understanding of how sellers' actions are likely to produce the environmental services that buyers want.

Because PES schemes are based around people, all chapters necessarily pay attention to the need for interactions with and between people. However, Chapter 7 is especially concerned with agreements between people. To ensure that any PES scheme functions well, contracts are required to bind people to their intentions. The role of intermediaries or brokers in any PES scheme is of particular importance in the contracting phase.

Expecting a PES scheme, once initiated, to continue to operate successfully over time is unrealistic. The environmental context, the people, the economic conditions, and a range of other factors that affect a PES scheme will inevitably change through time. This necessitates a process of continued assessment to be built into a PES scheme. Assessment is also important in providing confidence and trust to the parties involved in the PES scheme. Chapter 8 provides the framework for such an assessment task.

Chapter 9 makes the important point that this book is not intended to give an "ivory tower" perspective to designing and building a PES scheme. Experience has given the authors a very good idea of what can go wrong and what can be done to avoid problems or at least to deal with problems once they emerge. Every chapter points out pitfalls specific to the topic that chapter. To cap that off, the final chapter is devoted to the pitfalls but also to the promise of PES schemes. The intention is to forewarn and forearm the reader as well as give some suggestions and hopefully some inspirations for new PES ventures.

Finally, some specific "ways forward" are suggested in Chapter 10. The recommendations provided are designed especially for policy makers who have decided that PES schemes have potential as mechanisms for improving the protection of the environment. They relate especially to the establishment of an institutional setting in which PES schemes would be more likely to succeed.

1.4 The foundations

Environmental protection actions produce services that often have a characteristic that make them different from most other services that people enjoy. This characteristic is referred to as "open access" (see Box 1.1). The most striking impact of this characteristic is that environmental services are often not bought and sold in conventional markets.

Box 1.1 Open access

When an environmental service is "**open access**" it means that once it is available, it is impossible (or at least prohibitively costly) to stop people from using it without paying for it. Because of this characteristic, people can "**free-ride**" on any open-access service that is provided: people can use the service without paying for it in the hope that enough other people do pay to ensure its continued supply. Of course, with everyone thinking this way, no one pays and no open-access services will be produced commercially. Given these characteristics, it is therefore of little surprise that public goods such as some environmental services are not available through commercial channels.

The "open-access" characteristic is also known as "**nonexcludability**" and is one feature of environmental services that are "public" and "common pool." The difference between a service that is "common pool" and one that is "public" is that once provided a "**public**" environmental service is not diminished through use while a "**common pool**" environmental service, once used, will no longer be available. An example of an environmental service that is "public" is the knowledge that a species is protected from extinction. The extractive environmental services such as fishery and forestry products are examples of common pool resources. There is no commercial incentive to provide protection to endangered species and once existing stocks of common pool environmental services are exhausted (and there are strong commercial incentives for that to occur) there is no further profit incentive to restore them.

It is important to understand why the open-access characteristic occurs to be able to understand better how to develop a PES scheme that would fill the "gap" left by the commercial operation of markets where buyers and sellers get together (physically or virtually) to trade. Fundamentally, open access occurs because property rights over the environmental services produced cannot be well defined or defended. Property rights are the basis for any market trade. They are what are exchanged between buyers and sellers in any market transaction. If either party in an exchange cannot be assured that their rights to ownership are not secure, then the exchange is unlikely to progress. The defining and defense (and the exchange process itself) of property rights are themselves costly processes. These costs are known as "**transaction costs**." Where the complexities of defining, defending, and trading property rights are so high that they erode any possible gains by both buyers and sellers then the characteristic of open access or nonexcludability emerges as a barrier to the operation of markets. Transaction costs act as a barrier between buyer and seller: they can be so large that the gains to both parties from exchanging rights in a market are less than the transaction costs. That is when markets will not form and exchange between buyer and seller will not take place.

For example, if the costs of defining the rights to water quality and thereafter the costs of preventing those who do not pay from using the improved water quality are too high relative to the gains available from exchange, then an open-access situation arises. If the rights to the knowledge that an endangered species continues to survive cannot be adequately defined to prevent people who do not pay for the knowledge do not find out about it, then exchange is unlikely.

The existence of prohibitive transaction costs that prevent property rights being defined and defended is thus at the heart of the "failure" of markets to ensure the commercial provision of environmental services.

A market is a venue in which buyers and sellers can trade with each other. We often do not find open-access environmental services being bought and sold in markets: water quality improvements, storm surge protection, or endangered species protection are not available on the supermarket shelves or even online from Amazon.com.

Because no one is buying these services, actions to supply them such as revegetating stream upper catchments, preventing the clearing of coastal mangrove forests, the setting aside of habitat as protected reserves, or the control of invasive species create no revenues. Without revenues no one is likely to spend money paying for the costs of operating a business. Without revenues, businesses providing environmental service will be unprofitable. Hence, there is no commercial incentive for businesses to start up.

What can be done about this? It is not as if the environmental services are not valued by society. It is more likely that few if any people will pay for them to make the business of selling them profitable. The transaction costs of defining and defending property rights (see Box 1.1) are simply too high to allow the buying and selling of environmental services in conventional markets. The operation of Adam Smith's famous "invisible hand" in achieving the best (most efficient) use of available resources is thus impeded (see Box 1.2).

In conventional economic analysis, the existence of this "market failure" in the provision of environmental services gives reason for the implementation of some form of collective action. This is often observed in the form of a centralized government taking action to become a supplier of environmental protection, for instance as the owner and operator of nature protection reserves such as National Parks. It is also observed through the imposition of regulations (such as bans on clearing forested areas or pollution controls), the payment of subsidies (such as grants to reforest degraded catchments), and the levying of taxes (such as financial penalties for polluting or vegetation clearance).

Such interventions may be warranted from a conceptual perspective given the presence of the open-access characteristic but that alone does not justify government action. It must be demonstrated that the benefits created by the intervention exceed the costs. If they do not, then society is going backwards (in terms of overall well-being and economic efficiency) through the actions of government. This is an important test, especially when it is realized that government action itself involves transaction costs on top of all the other costs of providing services. These transaction costs include the operation of the bureaucracy (including the taxation mechanisms required to fund the intervention). Furthermore, there are concerns regarding the incentive of governments (politicians and bureaucrats) to achieve the environmental protection goals of the interventions.

In brief, the benefits of government action in providing environmental services may not warrant the costs, inclusive of the transaction costs.

It is this potential void between flawed government intervention and the inactivity of commercially motivated sellers when open-access conditions prevail that PES schemes have sought to operate.

Box 1.2 Adam Smith's "Invisible Hand" and "efficiency"

A product of the Scottish Enlightenment, Adam Smith, the so-called "father of economics" advanced his theory that the self-interest of individuals trading with each other in free and competitive markets would ensure the best use of resources across the whole of society. The logic is that trade, voluntarily undertaken by both buyer and seller, makes both parties better off. Such trading will continue until there are no further possible gains to be made. At that point, the resources of a society have been allocated to their highest value uses. Hence, people seeking to make themselves better off inadvertently make their whole society as well off as could be given the inherent scarcity of resources. Key to this process is the voluntary nature of exchange and competition between buyers and sellers—plus, of course, the presence of well-defined and defended property rights over the available resources.

Voluntary exchange comes about because of the incentives that buyers and sellers have to engage. Although exchange, buyers gain access to something that they want at a price that is lower than the value to them of the thing they buy (the buyer's "**willingness to pay**"). Sellers also have an incentive to engage. The price they are paid by the buyer is greater than the costs they have to incur in producing what they sell (the seller's "**willingness to accept**"). These costs are the value of the resources used in production in their next most valuable use. For instance, the seller's labor is a cost. The seller could have been employed in some other work and so the value of the seller's labor cost is the income foregone from not working at that alternative employment. This is the definition of an "**opportunity cost**" and it applies to all resources used.

In a voluntary exchange, the price agreed between the buyer and the seller is lower than the value enjoyed by the buyer but higher than the opportunity costs of the seller. That means the exchange ensures that the value enjoyed by the buyer is greater than the value that could have been generated if the resources were used in their next best use. The use of resources is said to be "more efficient" with the exchange than without it. This encapsulates the idea of "**economic efficiency**": if resources can be moved from one use to another so that one person is better off but no one is worse off then there is said to be an improvement in economic efficiency. The price that ensures economic efficiency is that which ensures the willingness to pay of buyers is equal to the willingness to accept of sellers.

1.5 What is a PES scheme?

A PES scheme is an alternative to the direct provision of environmental services by governments. A PES scheme aims to provide incentives for commercially motivated private sellers to produce environmental services that are paid for by buyers. In brief, a PES scheme should attempt to mimic the market process by providing incentives for buyers and sellers to exchange. By mimicking market processes, the advantages of market exchange in producing improvements in economic efficiency can be achieved.

PES schemes are designed explicitly to respond to the transaction costs that otherwise act as a barrier to the normal operation of markets when open-access environmental services are at stake. Taking a cue from conventional markets, PES schemes involve the operations of third party brokers that specialize in handling the exchange processes that give rise to the relatively high transaction costs. Just as real estate agents act to lower the transaction costs (such as information gathering and search costs) of people wanting to buy and sell houses, a PES scheme broker can act to facilitate the definition and defense of property rights over otherwise open-access environmental services by gathering information and reducing legal uncertainties. The idea is to lower the transaction costs of buyers and sellers to the point where exchange becomes worthwhile. PES brokers can be government agencies, nongovernment organizations and/or research organizations.

In this manner, PES schemes create "pseudo-markets" where conventional markets otherwise do not exist. They involve people who want an environmental service (buyers) and people who can provide the environmental service people want (sellers). Importantly, they also involve people who can bring buyers and sellers together by lowering transaction costs (brokers).

In line with the principles of Adam Smith's "Invisible Hand" of the market system (see Box 1.2), a PES scheme should, ideally, be voluntary and competitive. Buyers and sellers should not be forced to participate in the scheme. There should be multiple buyers and sellers engaged so that they experience competition and hence not hold any individual power to manipulate the price paid by buyers and received by sellers.

A PES scheme structured in this manner should ensure that buyers and sellers are all made better off by engaging with the scheme. With voluntary participation, anyone made worse off would not engage. With competition, the gains from exchange will not be concentrated in the hands of the few as would occur if there was a single seller or buyer who was protected from others entering the scheme. It should be noted that if the open-access characteristic of an environmental service is such that a PES scheme could only exist if the free riding incentive is counteracted by a government forcing buyers to pay, then it must be demonstrated that the gains from buyer participation are positive and greater than the costs.

Although a voluntary PES scheme so structured would ensure gains for buyers and sellers, it is also important to recognize the transaction costs incurred by brokers. These are the establishment and administration costs of operating the PES scheme that in turn allow the scheme to generate the gains to buyers and sellers. A PES scheme must be shown to create a net benefit for society as a whole, taking into account gains for buyers and sellers and the net costs of brokers. Otherwise, the society would be better off without it.

Similarly, the interaction between buyers and sellers in a PES scheme should ensure that additional environmental services are provided under the scheme than would be provided in its absence. If buyers find that their payments are not being rewarded with more environmental services, they will no longer trust the scheme and will cease to pay for it. This means that payments made to sellers should be conditional on delivery of contracted environmental services or actions. Where actions are contracted, the relationship between those actions and the final provision of environmental services must be well documented and accounted for in the PES scheme.

A key element in ensuring all of these features are embedded in a PES scheme is the transparent, timely, and reliable provision of information to all PES buyers, sellers, and brokers.

The key market element that a PES scheme should create is the price paid by buyers and accepted by sellers. A price per unit of environmental service output is paid by PES buyers. Most often, PES sellers are not paid directly for environmental services although where possible, this is advisable. Rather, a price is paid to sellers per unit of environmental management action used in the process of providing an environmental service. For example, a buyer may pay for improved water quality but a seller may be paid for planting trees in the catchment area. In calculating both prices, three elements of information are required:

1. The willingness to pay of the buyers
2. The willingness to accept of the sellers
3. The relationship between action inputs and environmental service outputs

With these three elements known, the efficient "market" price for buyers and sellers can be set by equating willingness to pay with willingness to accept, given the relationship between inputs and outputs.

1.6 Conclusions

PES schemes are designed to "fill gaps" left by conventional markets. They aim to bring potential buyers of environmental services together with potential sellers of environmental services (or management actions) through the actions of brokers who can act to lower transaction costs. The overall goal is to improve the well-being of people, a concept that is encapsulated in the concept of a more efficient use of available resources.

The process of designing and implementing a PES scheme is challenging, and hence costly. If it was not, there is little doubt that there would be no need for this book: myriads of independent privately motivated PES schemes would be already in place.

What this book seeks to do is to help the reader understand and then overcome the associated challenges. Information is key to this. The next three chapters of the book address the three core elements of information that are pivotal to any PES scheme. First, the biophysical relationships between sellers' actions and the environmental services that people want are explained. This provides the ecological backdrop to a PES scheme. Following that, the information regarding the respective wants of the two "sides" of a PES market—the buyers and the sellers—is detailed.

References and further readings

Publications of team members and collaborators of the Lao PES project:

Scheufele, G., 2016. Payments for environmental services. In: Bennett, J. (Ed.), Protecting the Environment, Privately. World Scientific Press, London.

Scheufele, G., Bennett, J., 2017. Can payments for ecosystem services schemes mimic markets? Ecosystem Services 23, 30–37.

Scheufele, G., Bennett, J., 2013. Research Report 1: Payments for Environmental Services: Concepts and Applications. Crawford School of Public Policy, The Australian National University, Canberra.

Publications of authors external to the project:

Barbier, E., Hanley, N., 2009. Pricing Nature: Cost-Benefit Analysis and Environmental Policy. Edward Elgar PublishingPublishing, Cheltenham.

Coggan, A., Whitten, S.M., Bennett, J., 2010. Influences of transaction costs in environmental policy. Ecological Economics 69, 1777–1784.

Frank, R., 2003. Microeconomics and Behavior. McGraw-Hill, Boston.

McCann, L., Colby, B., Easter, K.W., Kasterine, A., Kuperan, K.V., 2005. Transaction cost measurement for evaluating environmental policies. Ecological Economics 52, 527–542.

Mankiw, N.G., 1998. Principles of Economics. The Dryden Press, Harcourt Brace College Publishers, Fort Worth.

Mas-Colell, A., Whinston, M., Green, J., 1995. Microeconomic Theory. Oxford University Press, Oxford.

Tacconi, L., 2012. Redefining payments for environmental services. Ecological Economics 73, 29–36.

Tietenberg, T., Lewis, L., 2009. Environmental and Natural Resource Economics. Pearson International Edition, Boston.

Wunder, S., 2005. Payments for Environmental Services: Some Nuts and Bolts. CIFOR, Bogor, Indonesia. Occasional Paper.

Wunder, S., 2015. Revisiting the concept of payments for environmental services. Ecological Economics 117, 234–243.

Context

2

In Chapter 1, the basic principles underpinning Payments for Environmental Services (PES) schemes were outlined. They are the foundations on which an economically efficient PES scheme can be designed and built. In this chapter, the initial practical tasks of setting up a PES scheme are set out. This process commences with the recognition that there can be no single "formula" that will define a PES scheme. That is because the contexts of potential PES schemes are many and varied. Consequently, although the preceding chapter established the 3-core informational elements that underpin a PES scheme, the manner in which these elements are put together will vary from context to context.

Hence, this chapter is aimed at defining the characteristics of a potential PES scheme's context. The material presented in the chapters is structured so that for each characteristic, generic factors are initially considered. Then, the specifics of those characteristics as they apply to the case study featured in this book—the protection of endangered species in Lao People's Democratic Republic (PDR)—are set out. This is to demonstrate the practicality of the process of PES scheme design and establishment advocated in this volume.

2.1 Purpose

The first task faced by a PES scheme architect is to define the purpose of the scheme. In essence, this involves setting out a clear statement of the scheme's goal. At the core of any goal is a problem that requires a solution. The context of any PES scheme therefore must be shaped by the environmental service that is identified as potentially being at risk of excessive loss or having already been degraded, having no few prospects of repair or restoration.

Problems of this nature are usually only too apparent. Attention to them is frequently drawn by interest groups including adversely impacted local people as well as conservation lobby groups. For instance, local people may observe that the clearing of mangrove swamps along an adjacent coastline has been associated with increased impacts from high tides or storm surges. Downstream fishers may have correlated the loss of forest cover upstream with worsening water quality and fish catches. Local or even international NGOs may have conducted surveys of forest wildlife populations and found that the numbers of a specific species have fallen dramatically.

Buying and Selling the Environment. https://doi.org/10.1016/B978-0-12-816696-3.00002-8

Problem identification of this type usually involves the specification of the environmental asset and its associated service that would become the focus of a potential PES scheme. In the preceding examples, the mangrove swamp is the asset and the protection from storm surge is the service. The water quality is the asset and the fish catch is the service. The forest and the animals living therein are the assets and the knowledge of their continued existence is the service.

The goal of the proposed PES scheme should thus be directed at achieving the provision of the environmental service. That is what PES scheme buyers are interested in. Sellers are therefore directed toward either an expansion of current levels of provision or prevention of further reductions in provision, or both. Consequently, although the goal should focus on the service that is of interest to buyers, attention for the seller will necessarily also be focused on the asset that is the source of the service. More protection from storm surge will be provided by more mangrove forests. More fish will come from better water quality in the stream. Greater knowledge of species survival will come from having more forest and hence species.

It is also important at this early stage in the formulation of a PES scheme to consider the broader rationale for developing a scheme. This involves understanding the causes of the problem defined as the core of the goal.

Referring to the concepts set out in Chapter 1, problems associated with environmental services can arise because of their open-access characteristics and the associated high transaction costs with defining, defending, and trading property rights over those services. In such circumstances, there is an a priori case for intervention such as a PES scheme: without intervention, a conventional market will not be able to form to bring buyers and sellers together.

However, caution needs to be applied. The first reason for caution is the potential for earlier interventions to have caused the environmental service problem. Policies designed to achieve other purposes may well have had unintended perverse consequences for the environmental service. For instance, government subsidies paid to shrimp farmers to expand their operations to increase economic development may have caused coastal mangrove forests to be cleared to make way for shrimp ponds. State owned and operated forest operations upstream may be the source of deteriorating water quality downstream and biodiversity loss. Where environmental service problems arise because of past interventions, a PES scheme may not be well suited. The preferred approach may be to reconsider the design of the prior intervention to avoid the environmental side effects.

The second cause for caution relates to the extent of the transaction costs that are generated because of a PES scheme. It is true that problems with conventional market formation come about because of high transaction costs. However, it is also the case that the intervention alternatives to conventional markets, such as PES schemes, also incur transaction costs. The alternative to a problematic or nonexistent conventional market is not the "nirvana" of a costless PES intervention. The transaction costs of a PES scheme are predominantly the costs that incurred by the scheme broker. For a PES scheme to be appropriate, it must be demonstrated that the benefits so created for both buyers and sellers are in aggregate greater than the associated transaction costs. If this is not the case, then society is made worse off by the introduction of the PES scheme and the "status quo" situation should be maintained even though it involves environmental service losses.

The Lao PES scheme

The Lao PDR is a biodiversity hotspot. It is home to a range of wildlife species that are under threat from extinction. Most of these species are protected by customary laws and Lao statutory legislation. Some legal restrictions apply to the use of all wildlife species within and outside protected areas: Harvesting wildlife during gestation periods, wildlife with offspring, or by means of tools and methods that lead to increased depletion of wildlife is prohibited in principal. Other restrictions that are also binding within and outside protected areas apply only to species that are classified by Lao statutory legislation as being threatened with extinction (Category I and II species)[1]: Harvesting Category I species is prohibited in principle. Harvesting Category II species is generally prohibited but some exceptions are stipulated for subsistence use of local people. Buying or selling of both Category I and II species harvested in nature, including carcasses and animal parts, is prohibited in principal. Finally, some restrictions apply to all wildlife but within protected areas only: Harvesting or transporting any species within the total protection zone of a protected area is prohibited in principal.

Nevertheless, these wildlife species are under the threat from extinction because the existing law enforcement is largely ineffective in reducing poaching pressure. This means that the property rights over these species are largely defined but insufficiently defended.

Wildlife diversity in protected areas was identified as the asset that generates two environmental services buyers might be willing to purchase and sellers might be willing to take actions to increase their supply: the "knowledge of the continued existence of wildlife diversity" (hereafter called KNOWLEDGE) and (as a future option) the "opportunity to watch wildlife diversity" (hereafter called WATCHING).

Enforcing wildlife legislation in the Lao PDR poses a challenge: The wildlife habitats involved are vast, the budgets available for enforcement are insufficient, and the technical and legal expertise required is limited. In consequence, the risk of being caught and hence the expected costs of committing wildlife crimes such as illegal harvesting (poaching) are minimal. In contrast, the financial returns available from breaking the law are substantial. This attracts commercially motivated poachers from within and outside the Lao PDR.

Apart from commercially motivated poachers, wildlife crimes are committed—often unknowingly—by subsistence hunters. The rural population relies heavily on wildlife as a source of dietary protein. Yet, some people are simply unaware of

---

[1] *A list of wildlife categories is publicly available.*

the hunting restrictions stipulated by Lao legislation. They do not know that some of the species they hunt are legally protected, some of the methods they use are banned, certain hunting grounds are off-limits, or hunting is prohibited during gestation periods. A lack of information and awareness thus contributes to the loss of wildlife diversity.

2.2 Actions that produce environmental services

With the problem identified and a PES scheme approach deemed to be appropriate, the next part of the context to be established relates to the range of solutions that could be employed to deal with the problem. Just as the identification of the problem forms the base for linkages to PES scheme buyers, solutions are the links to potential sellers.

The definition of a "solution" deserves some clarification. In the previous section, the cause of the environmental service problems was traced back to markets failing to emerge for the environmental service that in turn was argued to be a function of the open-access characteristic. Hence, one solution to the problem may be the development of a regime of low transaction cost, excludable property rights for the environmental service at hand. This may well be the best approach to the problem and should be explored in the same manner as a policy maker should look for other policies that have the environmental problem as an unintended consequence. However, for the purposes of this exercise, what is of interest is the suite of actions that could be taken to remedy the shortfall of environmental services. In brief, what could be done to provide the environmental services that have proven problematic?

Hence, where mangrove forests have been removed to cause storm surge problems, the remedy may be a combination of replanting plus the construction of protective physical barriers. Upstream tree planting and erosion gully control structures may improve water quality. Biodiversity protection could be ensured through the control of predatory feral animals and invasive weeds.

To a large extent, the identification of protective or restorative actions is a matter for technical experts. People with knowledge and experience in disciplines such as wildlife ecology, landscape restoration, and pest control should be consulted. However, the knowledge and experience of local people and organizations should also be tapped, especially with an eye to a further context element: the identification of potential sellers of environmental services.

The Lao PES scheme

Defending the property rights over a range of endangered wildlife species was identified as a potential solution to minimizing further losses of wildlife diversity in protected areas and thereby to ensure the supply of the two environmental services KNOWLEDGE and WATCHING (see Section 2.1). This is achieved through wildlife protection actions performed in the context of a comprehensive antipoaching patrol scheme.[2]

    The antipoaching patrol scheme is at the core of the Lao PES scheme. Under the patrol scheme, patrol units are employed in a designated area to protect wildlife species from poaching. The designated area was divided into 1 km² grid cells, each marked by a central point using a GPS reference. The antipoaching patrol teams are instructed by a patrol manager to visit a series of assigned grid cells following specified patterns (for example, visiting the central points or following a zigzag line). Grid cells that are inaccessible to patrol teams and poachers are ignored. Both the grid cells patrolled and the sequence in which they are patrolled are randomized across patrol teams and time. This enables the patrol manager to respond to emerging threat hot spots, minimize predictability to poachers, detect changes in criminal activities, and enable effective wildlife and poaching monitoring. Knowledge of and access to patrol data is restricted by the patrol manager.

    To address safety concerns, patrol teams consist of five villagers, of which two or three are members of the village militia. The village militia consists of villagers who are appointed by the authorities to carry out security tasks over a specified period of time. They have the authority to carry weapons and, to some extent, enforce wildlife laws. All patrol team members were trained on patrol techniques and use of the equipment necessary to record field data, such as GPS units, maps, and patrol data forms. Employment was conditional on a successful completion of the patrol team training. One member of each patrol team acts as patrol team leader. Patrol team leaders were identified during the patrol team training.

    The patrol manager is responsible to plan the location and timing of the patrols and to identify "core zones" of high priority. This is achieved through a risk assessment that identifies "hot-spots," natural corridors, breeding sites, and entry/exit points. These "core zones" are subject to the most intense patrolling.

    The number of grid cells that constitute 1 day of patrolling varies to account for differences in effort. For example, a day is defined as patrolling 2—3 grid cells in the wet season and in steeper terrain compared to 3—4 grid cells in the dry season and flatter terrain.

    Patrol teams perform law enforcement and wildlife monitoring tasks. The tasks include the dismantling of snare lines and collection of snare wires, dismantling of poacher camps, detection and reporting of poaching activities, recording poaching incidents and evidence, confiscating poaching gear, confiscating poached animals and animal parts, issuing warnings to local poachers, apprehending poachers who are not Lao PDR citizen, and recording any direct and indirect sightings of key wildlife species. The patrol teams have to comply with an Environmental

---

[2] The core elements of the patrol scheme are based on a design developed by C. Vongkhamheng for the Nakai-Nam Theun National Protected Area but it was customized to this specific PES context. An expert survey was conducted to inform the scheme's design. The expert survey template is provided in Annex 1.

Code of Conduct developed under the Protected Area and Wildlife Project (PAWP) financed by the World Bank. The code ensures that patrolling in the designated area does not damage wildlife or their habitat. The code requires the patrol teams to minimize disturbance of wildlife, dismantle their camps before moving on and extinguish their cooking fires completely before continuing the patrol, minimize the cutting of vegetation and site clearing, extinguish cigarette stubs, and carry out any garbage (including cigarette stubs). They are not allowed to harvest any wildlife for food during an antipoaching patrol or make camp in ecologically fragile areas.[3]

The patrol scheme is embedded in the broader community by engaging the target villages as whole entities as components of the scheme. The participation of the complete villages is essential to secure support from villagers and village authorities. Broad community support and the associated peer group pressure was believed to increase the probability that the patrol targets would be achieved, patrol effort would be recorded reliably, and the risk of teams deliberately ignoring poaching activities would be reduced. Community participation focuses on increasing compliance rates with the statutory wildlife protection legislation. Additional voluntary agreements on wildlife protection were negotiated to cover legal "gray" areas associated with prohibitions regarding Category II wildlife species (see Section 2.1).

2.3 Spatial extent

A key element of any PES scheme context is the geographic area that is to be the focus of attention. Just as a conventional market involves sellers drawn from a specific area and provides for the needs of buyers in particular locales, so too does a PES scheme need to have spatial barriers defined.

In most circumstances, the nature of the environmental service and its associated problem goes a long way to define a spatial extent. If a problem is focused on an ecosystem, then the extent of the ecosystem can be the base for defining the market. For instance, a length of coastline that is exposed to storms but which has had its original mangrove stands removed could define a PES scheme. Upstream reaches of a river system and their subcatchment areas that are the sources of sediments and contaminants would also be a suitable base-level spatial definition as would the habitat areas of endangered species. These definitions rely on an identification of the location/s of the source of the problem and the environmental service involved.

However, it is also important to note the significance of buyers in this definition. A PES scheme can only work for an area where buyers are interested in and hold value for the environmental services provided. Hence, if mangroves have been removed from a 200 km stretch of coastline, but buyers can only be identified for

---

[3] Additionally, the patrol teams have to follow procedures developed to mitigate against damage or loss to cultural resources (Physical and Cultural Resources Chance-Find Procedure).

> ## Box 2.1 USE AND NONUSE VALUES
>
> Environmental services can be enjoyed by people who have a direct exposure or physical contact with an environmental asset. The benefits so enjoyed are known as "use values." For instance, a river may deliver fish catches to local people who either gain from catching the fish through direct consumption or by selling them to earn an income. Tourists may visit a scenic area to enjoy an encounter with an endangered species. Villagers may enjoy the protection of a mangrove forest "buffer" between their houses and the sea. Other benefits from an environmental asset may not require direct contact being made by a beneficiary. For example, a person can enjoy the knowledge that blue whales still exist without venturing onto the high seas for a direct encounter. This is a "nonuse value." In between these two types of values is another category known as "indirect use value." Here, the person experiencing the benefit of the environment can be located away from the asset itself but still be reliant on direct consumption of a derived service. For example, a person whose health is dependent on taking a pharmaceutical compound that was originally discovered in nature enjoys this type of value.

the lengths of coastlines that are adjacent to settlements, then a PES scheme would need to have its extent defined in terms of that buyer interest.

Similarly, political or bureaucratic boundaries may play a role in defining a scheme's geographic extent. This may be because of the need to conform to the legal conditions of a single jurisdiction. It may also arise because a government entity has a role to play in facilitating actions that could be taken to deal with the identified problem. Although specifying a PES scheme according to a bureaucratic boundary may have administrative advantages it may also trigger difficulties. For instance, a problematic river subcatchment may not be located in just one political jurisdiction. The additional transaction costs of extending a scheme across multiple jurisdictions then have to be weighed up against the benefits achieved from a better physically integrated scheme.

Although the "selling" side of a PES scheme will largely define a geographically limited scale, there is no such restriction on the "buying" side. This is particularly the case when the type of benefit buyers enjoys from an environmental service extend beyond "use values" to include "nonuse values" (see Box 2.1).

The spatial implication of nonuse values is that a PES "market" can extend globally on the buyer side. For example, where a PES scheme is aimed at protecting endangered species from extinction, people from far distant locations may have sufficient interest in the future of the species to hold an existence value and hence be potential buyers in the PES scheme. Similarly, visitors to an area may enjoy a use value. Where current and potential future visitors are resident in far distant locations, they may represent a source of PES scheme buyer interest.

The Lao PES scheme

The spatial extent of environmental service supply is defined by the boundary of the Phou Chomvoy Provincial Protected Area (PCPPA). It covers about 22,300 ha and provides habitat for a wide range of endangered wildlife species. Nineteen of these species were selected as target assets to be protected through wildlife protection actions (see Section 3.1). The PCPPA is part of the Northern Annamite Mountain Range. Its location on the Lao-Vietnamese border causes substantial poaching pressure, which poses a significant threat to wildlife diversity.

The scheme falls under the jurisdiction of one province (Bolikhamxay Province) and two districts (Khamkheuth District and Xaychumphone District). Government agencies involved in the supervision of the patrol scheme include the provincial and district Governor's Office, the Provincial Office of Forestry Inspection, the Provincial Agricultural and Forestry Office, and the two District Agricultural and Forestry Offices.

The wildlife assets that are protected under the scheme are located inside the PCPPA. The sellers who perform the actions that protect these assets are, however, located outside but adjacent to the PCPPA. That means that some of the scheme's impacts, such as alternative income opportunities through PES scheme engagement, reach beyond the boundary of the PCPPA. So strictly speaking, the spatial extent of supply is defined by the PCPPA boundary plus the villages of those who sell wildlife protection actions.

The spatial extent of environmental service demand, on the other hand, is neither restricted by the location of the problem nor the source of its solution. The environmental service KNOWLEDGE purchased by the buyers is a nonuse value. The buyers enjoy this service independent of their locations. The location of the buyers is irrelevant. The potential spatial extent of demand is therefore global. The environmental service WATCHING is a use value. Its enjoyment requires access to the PCPPA by buyers, which is currently restricted. If the access arrangements were to change, the PES scheme could be adjusted to produce the environmental service WATCHING as well. This would not change the spatial extent of the scheme. The spatial extent would still be global as buyers may come from all over the world to visit the PCPPA.

In summary, the geographic boundary of the Lao PES scheme is both spatially restricted (from a supply-side perspective) and unrestricted (from a demand-side perspective).

2.4 Time frame

The time element of a PES scheme may also require definition on both buyer and seller sides of the "market." Just as the spatial extent of a PES scheme may not be symmetrical across buyers and sellers, may there be differences in the temporal extent of PES schemes between buyers and sellers.

One factor that can be the cause of this asymmetry is the role played by the environment in responding to seller actions. Frequently in environmental management, the response of the ecosystem to restoration actions takes a considerable period of time. Planting trees will yield sediment reduction responses over years and even

The payment collection systems are compulsory insofar that all tourists and all electricity-using residents would have to pay, including those who do not value the environmental service KNOWLEDGE. This payment collection system could easily be expanded to the environmental service WATCHING. It might be argued that the free-riding problem associated with this environmental service could be overcome without "forcing" a payment on people who have no interest in the environmental service. This might be achieved by introducing an entry fee to the PCPPA instead of the visa entry fee and electricity surcharge. However, introducing separate payment collection systems for different environmental services would increase the transactions costs of the scheme. This is especially the case if the scheme was to be expanded across all protected areas of the Lao PDR.

Until the sustainable funding mechanism described previously is in place, seed funding provided by World Bank under the PAWP is being used. Hence, the current buyers in the PES scheme can be regarded as the international community, represented by the governments that provide funding to the operations of the World Bank (see Section 2.7).

2.6 Potential sellers

Sellers engaged in a PES scheme can provide environmental services directly or indirectly via the conduct of actions that eventually lead to the generation of more environmental services. The distinction between direct and indirect provision often arises because of the spatial or temporal "disconnect" between seller actions and environmental service supply. For instance, upstream tree planters may be far distant from downstream water users. Their actions may also take many years before they have measurable impacts on water quality. Even people reestablishing a coastal mangrove forest immediately adjacent to a village threatened by storm surges will face a time lag between action and outcome. In contrast, some sellers may be able to offer am environmental service immediately. This can be the case when the service being offered is the protection of an environmental asset. The service being sold in such cases is a continuation of provision.

In contrast to the buyer identification task, particularly nonuse value buyers, finding direct environmental service sellers is likely to be straightforward. This is because they are most commonly located adjacent to the site where the services are being provided or could be provided. Sellers located near the source of the environmental services are more likely to be more competitive (that is have lower costs) because they have superior knowledge of the sites involved and lower costs of transporting themselves to the sites. Indirect sellers, that is, sellers of actions rather than environmental services, may be more difficult to identify, particularly where there is a geographic "disconnect" between sellers and buyers.

It is not always the case that local sellers will be clearly most able to act. Four key factors influence the potential of PES scheme sellers to engage. First, they must have the necessary skills/knowledge to perform the tasks involved. Second, they need to have the time available to perform the task, given other calls on their time. Third, an interest in performing the tasks may be important. Finally, in some cases, access to appropriate equipment that is complementary to the sellers' labor may be required.

Local people with local knowledge of an environmental issue and its management may be appropriate sellers. However, the skills necessary to restore or protect an ecosystem may involve complexities beyond the initial capacity of locals. Specialist sellers may therefore be required. Alternatively, local people may have their capacities enhanced through training programs to the point where they are able to become sellers in the intermediate and longer terms.

Time is also a critical factor in determining seller availability. Of course all potential sellers are time constrained and all potential sellers have other things to do with their time apart from providing environmental services. Those alternatives may include leisure activities or they may be other forms of earning an income. If local people have few other income earning opportunities and what they are giving up by becoming sellers of environmental services is their leisure time, then they have low "opportunity costs" and so are more likely to be potential sellers. If, however, there are high paid jobs available in the local area, locals are less likely to engage with a PES scheme.

An interest in the environmental services being provided can also be useful in recruiting potential sellers. For instance, if local people are beneficiaries of a PES scheme, they are more likely to want to become a seller. This is because their costs of being a seller are offset to at least some extent by the benefits they receive not only from the income received but also by their enjoyment of the environmental services they provide. For instance, person living on the seafront is more likely to want to be engaged as a seller of restored mangrove forests. Similarly, someone with a love of a particular endangered species may be particularly keen to engage with a PES scheme as a seller.

Equipment can also be a limiting factor in establishing as a seller. For instance, local people may not have access to earthmoving equipment needed to construct gully erosion structures in a degraded catchment. Where such structures are important complements to tree planting operations, they may need to be contracted to more distant sellers while the local people could be engaged as tree planters and managers. Note that equipment and labor can also be substitutes in so far as a large labor force can perform the same tasks as heavy machinery. The mix of the two is dependent on their relative costs.

Sellers may be individuals or organized in groups. In some circumstances, whole communities may become sellers. Providing opportunities for the broader community to engage in a PES scheme is often essential to secure community support (see Section 5.1). Engaging only a subset of community members may jeopardize a scheme's social cohesion and sustainability. However, in the absence of effective social controls engaging whole communities as sellers may invite extensive free riding. Unchecked, free riding may render a scheme ineffective in delivering the desired environmental service.

The Lao PES scheme

The source of potential sellers are eight villages in the Bolikhamxay Province (seven are located in the Khamkheuth District and one in the Xaychumphone District). The Lao PES scheme engages the villages as whole entities as well as patrol teams comprising individual members of the villages. Both the individuals and the villages are indirect sellers who perform wildlife protection actions (see Section 2.2) instead of providing the environmental services directly.

The eight villages are home to 1029 households. In five villages the households are a mixture of ethnic groups, while three villages are inhabited exclusively by a single ethnic group. Village accessibility differs across villages. Two villages are located near the main road to Vietnam and are accessible all-season. The other six villages are only accessible via unpaved roads. This makes transportation during the wet season difficult (five villages) or impossible (one village).

The education level (including literacy rates) of the villagers is generally low. Villagers are mostly subsistence farmers with limited employment and income opportunities outside the agricultural sector. Cash income varies within and across villages. The differences across villages may be explained by their relative distance to the main road and so access to markets. They signal potential differences in the villagers' opportunity costs of time.

The villages are located near the boundary of the PCPPA. This minimizes the costs of the patrol teams to transport themselves to the area and back home. It also ensures that the villagers are connected with the area, are familiar with its environment, and so have had already some of the skills required to perform antipoaching patrols. Yet, the villagers had neither the expertise nor the access to markets to purchase the appropriate patrol equipment. The patrol equipment had to be provided by the scheme administrators. Some of the equipment remains the property of the scheme administrators, including the GPS units, cameras with in-built GPS function, radios, maps, compasses, binoculars, backpacks, flysheets, hammocks, and first aid kits. Other equipment items were part of the payments to the patrol teams, including field clothing (same color for all patrol teams and marked as "village patrol"), hats, boots, antileech socks, and mosquito repellent and remain the property of the patrol teams/individual patrollers.

The villagers depend on the use of wildlife and other forest resources. Even though the villagers have a self-serving interest in protecting these resources, their knowledge of the use restrictions stipulated by Lao legislation is, at best, limited. Knowledge of and compliance with these restrictions would support a sustainable use of wildlife and other forest resources in the PCPPA.

Villagers are generally concerned by the activities of poachers who come from outside their villages. As the villagers view these poachers as a threat to their livelihoods, they showed a great interest in engaging in wildlife protection actions, and specifically, in antipoaching patrolling. Hence, they are not only potential sellers but also beneficiaries of the PES scheme.

2.7 Potential brokers

In Chapter 1 a PES scheme broker was defined as an entity (person or organization) that bears some of the transaction costs involved in a scheme with the intent of facilitating exchange between buyers and sellers. Brokers, as specialists, have the potential to enjoy economies of scale in bearing transaction costs. By bringing down the transaction costs, brokers lower the barriers that otherwise separate potential buyers and sellers.

A key driver of transaction costs in a PES scheme or indeed in any market is uncertainty. Buyers and sellers are frequently separated because they are unsure about each other and the prospects of securing what they each want from any exchange. There may well be a time gap between sellers taking action to restore an environmental asset and the buyer receiving the associated service. Buyers and sellers may be located distant from each other, particularly when nonuse values are involved. Buyers and sellers are unlikely to know each other personally. All of these factors cause uncertainty. For the seller, there is uncertainty as to whether or not buyers will pay either now or into the future. For the buyer, there is uncertainty as to the seller's commitment to provide the desired environmental service so that it is available into the future.

Hence, brokers face the task of generating knowledge and hence trust among buyers and sellers. This can take the form of negotiating agreements among buyers and sellers to participate.

Underpinning such agreements are the usual forms of "trust building" that are at the base of any commercial agreement, notably elements of the legal system. For instance, agreements are established using contracts, written or verbal. These are legally binding statements of expectations that are backed, in turn, by the legal system that provides for the enforcement of penalties for breaches. In applications where there is a fundamental weakness in the underlying legal system, a PES scheme would need to deal with the associated additional transaction costs (brought about because of the additional uncertainty). Measure required could involve additional contractual clauses or the use of cultural sanctions that are beyond "black letter" law.

Brokers can play a key role in the provision of a range of types of information. The identities of potential buyers and sellers can be sourced and made available so that negotiations can be initiated.

Another form of information that can be generated and made known by brokers relates to the relationships between the actions taken by sellers and the environmental services desired by buyers. For instance, a prospective buyer is likely to want to know what impacts on water quality an investment in catchment tree planting is likely to have. Such information is critical for an assessment of value for money. Similarly, a seller would like to know the efficacy of alternative actions in the delivery of biodiversity protection: Is feral animal control more effective than habitat protection?

Such information may be sufficient for buyers and sellers to commence negotiations independently. However, a broker may be required to facilitate those negotiations either in person or virtually through their position as a mutually trusted independent party.

Where high transaction costs remain in the coordination of buyers and sellers, brokers may be able to intervene. For instance, if an environmental service is going to be enjoyed by a large number of heterogeneous people, coordination to prevent free riding is likely to involve high costs. Governments with the right to tax their populations can act in these circumstances to act as a buyer in a PES scheme, effectively taking on the role of a broker to lower the costs of coordinating individual buyer payments. On the other side of the PES "market," coordination among sellers may also be required to produce environmental services cost-effectively. This may extend beyond coordinating disparate sellers across different areas of operation in which they each have a comparative advantage (temporally, spatially, and skill sets) but it may also be required where the provision of an environmental service will only be achieved if sellers collaborate. For instance, species that are being protected under a PES scheme may be migratory and different sites may need management to ensure species survival. Seller coordination may also be a priority in settings where the education levels of potential local sellers are low.

Once negotiations have delivered a workable agreement between buyers and sellers brokers may have a role in monitoring and enforcing the agreement. Part of that role can be acting as a clearinghouse for funds: funds are collected by the broker from buyers and then distributed to sellers. This role can extend to the mediation of disputes.

Trust in the broker is clearly an important part of ensuring its role as a force in lowering transaction costs. This trust has to be between the broker and buyers and sellers. The selection of a broker for a PES scheme must therefore focus on the trustworthiness of a potential broker and their familiarity with both buyer and seller groups.

Other skills are needed by a broker. The capacity to understand the ecological system, and ability to deliver information relevant to the operation of a PES scheme are important.

Given that the skill set required to broker a PES scheme can be extensive, it is likely that in more complex PES schemes where the gap between prospective buyers and sellers is relatively wide, specialization in the tasks performed may be warranted. For example, a local NGO may be most trusted among prospective sellers to coordinate their activities and negotiate on their behalf. A government agency may need to take the role of buyers' agent if free riding is manifest. Neither may have the skills necessary to establish a quantitative relationship between sellers' actions and the environmental service being bought by buyers. Those skills are most likely available through dedicated research organizations.

When such a range of skills and hence brokers is required, an overarching coordinator is likely to be necessary. The implications for the transaction costs of the scheme are clear, and it is likely that such a complex scheme would only be economically efficient—that is, its benefits are greater than its costs—when the environmental services supplied are particularly valuable.

The Lao PES scheme

The origin of the Lao PES scheme largely dictated the fundamental structure of the brokers involved. The scheme originated within a research project aimed at investigating the practicalities of initiating a PES scheme in Lao PDR. The research work was funded by the Australian Center for International Agricultural Research (ACIAR) and was established under an agreement between the governments of Australia and Lao PDR. The Lao Government (GoL) therefore was explicitly involved from the outset. So too were the research organizations: The Australian National University (ANU), The National University of Laos (NUoL), and the University of Western Australia (UWA).

As the research work progressed, "brokering" specializations emerged within the initial project partners and in other organizations recruited to assist. Negotiations with prospective sellers were conducted by NUoL in collaboration with provincial and district staff of the Ministry of Agriculture and Forestry and the Governor's Offices. The Luxembourg Agency for Development Cooperation (LuxDev) assisted in the formulation of the processes to distribute payments to Village Development Funds, including the dual use of the funds they had already established.

Numerous attempts to engage buyers were made by ANU and NUoL in collaboration with the Environmental Protection Fund (EPF) within the GoL. Talks were held with a number of mining and hydroelectricity companies operating in Laos without success. Seed funding to establish the scheme and to see it operate for an initial 3-year period was finally sourced from the World Bank. As such, the buyers can be viewed as the international community, represented by the governments that fund the World Bank's operations.

The biophysical knowledge "brokering" was performed by a local Lao NGO, the Wildlife Conservation Association (WCA), and UWA. Site observations and ground trothing were conducted by WCA while UWA constructed the models needed to relate seller inputs and biodiversity outcomes. The Wildlife Conservation Society assisted through valuable discussions and an ongoing exchange of information. The antipoaching patrol scheme was developed by the ANU and WCA with support from the World Wildlife Fund.

The economic analysis necessary to estimate sellers' and buyers' costs and benefits and the prices to be paid and received was performed by ANU in conjunction with NUoL. The ANU team was responsible for the oversight of the research phase of the scheme. The Australian taxpayers, through ACIAR, funded the research costs that underpinned the establishment of the scheme.

Once the PES scheme structure was established and funds received, the EPF became a clearinghouse overseeing the day-to-day operations of a newly appointed team of scheme managers set up within the NUoL. The managers were responsible for the contracting of sellers and the provision of equipment. The EPF also became the final arbiter for dispute resolution.

2.8 Challenges and limitations

Environmental services may not be provided through the operation of conventional markets. This is largely because the transaction costs involved in bringing buyers and sellers together to exchange property rights are simply too high to be overcome by the prospective benefits available to those buyers and sellers. However, this should not be taken to be the sole justification for a PES scheme. This chapter has demonstrated that PES schemes that are designed to overcome the transaction costs of conventional markets are themselves subject to transaction costs. As the complexity of any PES scheme grows with greater barriers between buyers and sellers, so too do the costs of establishing and operating a PES scheme. These costs pose the biggest challenge to any PES scheme.

There is little doubt that potential buyers and sellers can be found for most environmental services that are neglected in conventional markets. However, complexities are everywhere. Free riding on the buyer side of the PES scheme equation is a key barrier. Where large numbers of diverse environmental service beneficiaries are evident, there is little prospect for forces such as peer group pressure or threat of group exclusion to cause free riding to be overcome. Then only collective action among buyers is likely to succeed. This may come through NGOs but most frequently, environmental NGOs find greater rewards in lobbying governments to secure supply rather than collecting private funds from buyers to ensure provision. Similarly, businesses with environmental connections such as agricultural, mining, and hydropower companies may be under no obligation to buy environmental services, even when their shareholders seek "corporate social responsibility." Frequently, corporations do face environmental care obligations that are imposed by government. Taking on further environmental purchases can therefore be seen as "doubling up." In that manner, government regulations regarding the environment can "crowd-out" the prospects of corporate sponsorship of PES schemes.

The up-shot of all this on the buyer side is that it is likely that governments will be involved as surrogate buyers. This limits PES schemes in a number of ways. First, it places them in the sphere of politics rather than the market. That opens them to the uncertainties of political terms and lobbying. Second, it removes the "voluntary" component of a PES scheme. Some taxpayers will be funding schemes they do not value while others will want more than they are receiving under the scheme. Third, the flexibility of a scheme will likely be severely restricted under the bureaucratic regime of government involvement.

Neither is the seller-side clear-cut. The context of a scheme may deliver vagaries of location and extent. Boundaries are unlikely to be exact. The relevant time frame will also be difficult to match, especially between buyers and sellers.

Multiple environmental services pose particular challenges. The protection or restoration of an ecosystem may provide a range of different benefits to a range of different people. That implies a PES scheme with differential prices across buyers. Furthermore, there may be different sellers providing different tasks, again implying differential payment regimes.

With so many potential complexities, the purist may seek a "perfect" PES scheme designed to deal with each specific aspect. The reality is that such a scheme will undoubtedly face overwhelming transaction costs. Just as "perfect knowledge" and "perfect markets" do not exist in reality, "perfect PES schemes" are also imaginary. The most any real PES scheme can aim for is an increase in the well-being of people through an improvement in the provision of environmental services. Compromises are always necessary in the process of designing and implementing PES schemes. Trade-offs between sophistication and practicality will be required to ensure that the transaction costs do not become too onerous. In the end, an evaluation of the costs and benefits of a PES scheme, relative to the situation without one, must be able to demonstrate an improvement in net social well-being.

References and further readings

Publications of team members and collaborators of the Lao PES project:

Hay, E., Kragt, M., Renten, M., Vongkhamheng, C., 2017. Research Report 11: Modelling the Effects of Anti-poaching Patrols on Wildlife Diversity in the Phou Chomvoy Provincial Protected Area. Crawford School of Public Policy, The Australian National University, Canberra.

Scheufele G., Bennett J.. Valuing Biodiversity Protection: Payments for Environmental Services Schemes in Lao PDR, Environment and Development Economics. May 2019, online first.

Scheufele, G., Bennett, J., 2018. Costing biodiversity protection: payments for environmental services schemes in Lao PDR. Journal of Environmental Economics and Policy 7, 386—402.

Scheufele, G., Bennett, J., 2017. Can payments for ecosystem services schemes mimic markets? Ecosystem Services 23, 30—37.

Scheufele, G., Bennett, J., 2017. Research Report 13: Valuing Biodiversity Protection: Payments for Environmental Services Schemes in Lao PDR. Crawford School of Public Policy, Australian National University, Canberra.

Scheufele, G., Bennett, J., 2017. Research Report 16: Costing Biodiversity Protection: Payments for Environmental Services Schemes in Lao PDR. Crawford School of Public Policy, Australian National University, Canberra.

Scheufele, G., Bennett, J., 2013. Research Report 1: Payments for Environmental Services: Concepts and Applications. Crawford School of Public Policy, The Australian National University, Canberra.

Scheufele, G., Bennett, J., Kragt, M., Renten, M., 2014. Research Report 3: Development of a 'virtual' PES Scheme for the Nam Ngum River Basin. Crawford School of Public Policy, The Australian National University, Canberra.

Scheufele, G., Smith, H., Tsechalicha, X., 2015. Research Report 4: The Legal Foundations of Payments for Environmental Services in the Lao PDR. Crawford School of Public Policy, The Australian National University, Canberra.

Scheufele, G., Vongkhamheng, C., Kyophilavong, P., Tsechalicha, X., Bennett, J., Burton, M., 2016. Research Report 9: Providing Incentives for Biodiversity Protection: Anti-poaching Patrolling in the Phou Chomvoy Provincial Protected Area. Crawford School of Public Policy, Australian National University, Canberra.

Tsechalicha, X., 2017. Research Report 14: Engaging Communities in a Payments for Environmental Services Scheme for the Phou Chomvoy Provincial Protected Area. Crawford School of Public Policy, Australian National University, Canberra.

Tsechalicha, X., Pangxang, Y., Phoyduangsy, S., Kyophilavong, P., 2014. Research Report 5: The Environmental, Economic and Social Conditions of the Nam Mouane — Nam Gnouang Catchment. Crawford School of Public Policy, The Australian National University, Canberra.

Vongkhamheng, C., 2015. Research Report 8: Phou Chomvoy Provincial Protected Area: A Biodiversity Baseline Assessment. Crawford School of Public Policy, Australian National University, Canberra.

Publications of authors external to the project:

Barbier, E., Hanley, N., 2009. Pricing Nature: Cost-Benefit Analysis and Environmental Policy. Edward Elgar PublishingPublishing, Cheltenham.

Tietenberg, T., Lewis, L., 2009. Environmental and Natural Resource Economics. Pearson International Edition, Boston.

Linking inputs with outputs

Any Payments for Environmental Services (PES) scheme requires the bringing together of those who want environmental services with those who have the capacity to provide them. The integration process that this involves is conducted within a biophysical context. Understanding the relationships existing between the inputs provided by potential suppliers of environmental services and the outputs that are wanted by potential buyers is therefore a critical step to take in designing a PES scheme. This is a relationship that exists in the biophysical environment and thus requires knowledge of the environmental context. In effect, it involves the development of an understanding of the effectiveness of environmental management actions in producing environmental services. This relationship is known as a "production function" (see Box 3.1).

Importantly, the relationship between inputs and outputs has to be established using comparable units of measurement. Data on inputs must be compatible with data on outputs. Hence, a critical issue in PES scheme design is to ensure that the information sets collected for the input side of the scheme (what sellers do) are compatible with the data relating to outputs (what buyers want). Specifically, this means that the buyers' willingness to pay data needs to be able to be matched with the sellers' willingness to accept data. Only then can an environmental service price be set for both buyers and sellers. The stumbling block to this process is that in many contexts, the seller information relates to willingness to accept to perform environmental

---

Box 3.1 Production functions

A **production function** sets out the relationship between inputs into a production process and the outcomes that result. For instance, the production of wheat involves inputs of land, water, fertilizer, machinery, and labor. The production of environmental services similarly requires inputs of various types. More inputs generally result in more outputs so that the relationship is positive. More fertilizer means more wheat. More people and tree seedlings mean better water quality downstream. How extra output can be generated by adding inputs is explained by the production function. In general, past a certain amount of inputs, as more inputs are applied, the extent of additional outputs generated starts to decrease. In other words, the "productivity" of an input starts to fall off. For example, a farmer adding more fertilizer to their crop will find that the extra yield created rises but at a decreasing rate. Eventually, yields may even decline with more fertilizer being used. Similarly, water quality improvements will be strong when the first trees are planted and have become established, but once a certain density of trees has been achieved in an area, more trees in the same location will not improve downstream water quality any more. This phenomenon is known as diminishing marginal product.

---

management actions (input space) while the buyer information involves willingness to pay for environmental service outcomes (output space). What is required is a conversion from environmental management actions to environmental service outcomes or vice versa. That enables sellers to be paid for their inputs while buyers are charged for the outputs they buy.

The goal of this chapter is to set out the process used to estimate production functions for PES schemes. This process is fundamentally one that involves the construction of a "bioeconomic" model. Hence, the next section sets out the fundamentals of bioeconomic modeling. The process as applied in the Lao PES scheme context is detailed. Such production functions allow the conversion of units of inputs into units of outputs or vice versa. How such bioeconomic models can be estimated is outlined in Section 3.1 of the chapter, again with illustrations from the Lao application. Finally, challenges and limitations of the approach are detailed. These are especially important in application contexts where existing knowledge of biophysical relationships is limited, a circumstance that is very common in developing countries where biodiversity is especially at risk.

3.1 Bioeconomic modeling

A model is a simplified representation of a complex system that is delineated by spatial and temporal boundaries. Typically, a model is based on a range of simplifying assumptions and focuses on the system's core parameters and relationships, which are specified by a set of equations. Reducing system complexity facilitates an analysis of present, a recreation of past, and a prediction of future system processes and states.

A bioeconomic model combines biophysical with economic parameters and relationships. In the context of a PES scheme, economic parameters specify drivers of environmental degradation and management actions to increase (or prevent a further decrease) in the provision of an environmental service. Such a model can be used to predict the effectiveness of management actions (inputs) in producing environmental service outcomes (outputs). Therefore, model predictions can be used to inform production functions.

The model type and design must be suitable to estimate the production functions relevant for a PES scheme. Models are usually customized to a specific temporal and spatial scale and developed for a specific purpose. Predictions outside the model specifications are likely to be poor. It is also important to acknowledge that every model embodies trade-offs between realism, precision, and generality. Which of these three characteristics dominates depends on the specific purpose of the model.

The model development process can be disaggregated into five steps. Each step requires the analyst to answer a range of questions.

Step 1: Definition of model purpose:

What are the environmental services buyers are willing to purchase and the environmental management actions that sellers are willing to undertake to produce the environmental services? What are the threats that prevent or reduce the supply of these environmental services? What management actions could remove or reduce these threats and/or their impacts?

Step 2: Model development:

What are the spatial and temporal specifications of the model? What are the relevant parameters and relationships that need to be simulated to represent the ecosystem, its core processes, and the effects of threats and management actions designed to reduce these threats? How can the parameters be quantified? What equations can be used to simulate the relationships? Which of the parameters are relevant to establish a quantitative link between environmental management actions (inputs) and environmental service outcomes (outputs)?

Step 3: Data collection:

What data are required to populate the conceptual model? What data are available? How can any data deficiencies be managed?

Step 4: Model estimation:

Are the collected data compatible with the model specifications? What units of measurement are suitable to quantify the parameters?

Step 5: Sensitivity analysis:

Does the model produce essentially the same results under constant conditions (reliability)? Does the model produce results that are plausible and conform to real world measurements (validity)? To what extent do results change over ranges of parameter values (robustness)?

The Lao PES scheme

Step 1: Definition of model purpose:

Wildlife diversity in the Phou Chomvoy Provincial Protected Area (PCPPA) is under threat from poaching. A comprehensive antipoaching patrol scheme was identified as a potential solution to minimizing further losses of wildlife diversity (Chapter 2). Nineteen species were selected as target assets to be protected through the patrol scheme to ensure the supply of two environmental services, KNOWLEDGE and WATCHING:

1. *Rufous-necked hornbill*
2. *Asiatic black bear*
3. *Tiger*
4. *Clouded leopard*
5. *Sambar*
6. *Southern Serow*
7. *Large-antlered muntjac*
8. *Saola*
9. *Douc langur*
10. *Northern white-cheeked gibbon*

11. *Sunda pangolin*
12. *Chinese pangolin*
13. *Pygmy slow loris*
14. *Bengal slow loris*
15. *Large-spotted civet*
16. *Owston's civet*
17. *Annamite striped rabbit*
18. *Northern pig-tailed macaque*
19. *Stump-tailed macaque*

The selected species are targeted by poachers and are of national and global conservation significance. They have been categorized by the International Union for Conservation of Nature (IUCN) as endangered and critically endangered and/ or are listed in the PCPPA management plan. Their likely presence in the PCPPA was assessed through expert opinion[1] and a wildlife survey.[2]

The main purpose of the bioeconomic model is to predict the effect of the anti-poaching patrolling activities (inputs) on the population size of each of the target species (outputs). This was achieved through the development of a stochastic dynamic population model.

A dynamic model specification simulates how a system's parameter values change over time. A population model simulates how parameter values relevant for population dynamics change over time. Consequently, any population model is inherently dynamic.

Population data on wildlife within the PCPPA were poor and/or limited. That is, the data, and therefore the parameter values included in the model, are characterized by a high degree of risk and uncertainty. A stochastic model specification addressed these limitations to some extent by including probability distributions into the calculations. The model was developed in the programming language R.[3] Using a free software removes access barriers in contexts, where funds to purchase commercial software are limited.

Step 2: Model development:

The model was developed to predict the population size of each of the target species. It was defined by spatial and temporal specifications as well as a range of parameters and functions that allowed the simulation of core processes.

Spatial specifications. A map of the boundary of the PCPPA was used to specify the spatial delineation of the model. The area and the boundary of the PCPPA were plotted in EXCEL to create a "boundary map." Each grid cell within (outside) the boundary map represents a spreadsheet cell with the assigned value of one (zero). Animal movements as well as poaching and patrol activities were

---

[1] The expert survey used to inform the development of the model is presented in annex 1.
[2] The wildlife survey was commissioned by the project 'Effective Implementation of Payments for Environmental Services schemes in Lao PDR'.
[3] https://www.r-project.org/.

restricted to the area of the PCPPA. This was achieved by linking the boundary map to all of the datasets and maps that were used in R. The area of the PCPPA was partitioned into 1 km² grid cells (in accordance with the design of the patrolling scheme set out in Section 2.2). Partitioning the area of the PCPPA enabled the assignment of parameter values at the grid cell level.

Temporal specifications. The model was designed to simulate core processes on a monthly basis (monthly "time step"). The model was designed to simulate these processes over any specified number of months.

Habitat quality. This parameter was included in the model to provide information required for defining the home range of the target species. Each grid cell was assigned a habitat quality value between zero and one. The value assigned to each cell was calculated based on its forest cover and relative distance to waterbodies using an asymptotic function. This function ensured that the habitat quality value of a cell increases with decreasing distance to a waterbody and/or greater forest cover. Information to estimate forest cover as well as size and location of waterbodies was derived from satellite images and topographical maps. Water bodies located outside the area but presumed to affect the habitat quality of cells located inside the area were included in the model.

Accessibility. This parameter was included in the model to capture the impacts of poaching and patrolling activities in the PCPPA. It was assumed that the accessibility of a cell depends of its relative distance to villages and/or roads. Information on accessibility was included in the model by assigning each grid cell with an accessibility value between zero and one. The value of each cell represents the probability of poachers and patrols moving through the cell. Cells covering village areas and roads were assigned fixed values. The value of each other cell was calculated based on its relative distance to roads and villages using an asymptotic function. This function ensures that the accessibility value of each decreases with an increasing distance to a village or road. The function contains a constant that determines the degree of the overall accessibility. This constant can be readily adjusted if more data indicating a different degree of accessibility (for example, data on actual poacher movements) become available. Information on the location of roads and villages was derived from satellite images and topographical maps. Roads and villages located outside the area but presumed to affect the habitat accessibility of cells located inside the area were included in the model. The so created "accessibility map" was combined with the "boundary map" to guarantee that the model simulates poacher and patrol movements exclusively within the area of the PCPPA. Assigning accessibility values to individual grid cells facilitates the modeling of poacher and patrol movements. Movements can be restricted to those grid cells that are assumed accessible to both poachers and patrols. Adjusting this assumption in the model is straightforward. Poacher target areas can be changed readily if new data become available.

The targeted wildlife species and their vulnerability toward poaching activities were characterized by the following parameters and relationships:

Initial population size. This parameter was included in the model to establish a starting point for the simulation of population dynamics. It was defined as the current population size of the target species present within the PCPPA (initial state). The assigned parameter values are species-specific.

Reproduction rate. The parameter was included in the model to simulate the number of additional offspring produced at each time step. It was defined as the average number of offspring a mature female produces per year. The assigned parameter values are species-specific. It was assumed that 50% of all animals are females. The model allowed each mature female to produce offspring at any time step. The number of offspring each female produced was chosen at random by drawing from a Poisson distribution of the numbers that are possible.

Generation length. This parameter was included to determine the number of mature females at any given time step. It was defined as the number of years since birth until reproductive maturity. The initial age of each animal was chosen at random by drawing from a uniform distribution defined between zero and double the age of reproductive maturity. The age of each animal was tracked at each time step during the model simulations. The assigned parameter values are species-specific.

Group size. This parameter was included to determine the size of family groups. It was defined as the number of animals living together. The assigned parameter values are species-specific. Species that typically live solitarily were assigned a group size of one. The initial group size was calculated by dividing initial population size by group size (rounded to the nearest integer).

Animal range. This parameter was included to specify the territory occupied by each animal group. Each group (and solitary animal) was initially assigned a territory delineated by a species-specific number of grid cells. The size of the territory (and hence the number of grid cells) reflects the range of each species. Grid cells were drawn at random from all grid cells located within the PCPPA. The probability of a grid cell being chosen was weighted by its habitat value. Grid cells with a higher habitat value were more likely to be chosen than those with a lower value. It was assumed that the occupied territory remains the same over the lifespan of each group (or solitary animal). Offspring produced after model initialization were assumed to occupy a vacant territory chosen at random for all unoccupied grid cells within the PCPPA. Groups (and solitary animals) were assumed to move within their territories with an equal probability of occupying any grid cell at any time. Dispersal and recruitment beyond the boundary of the PCPPA was assumed to be non-existent. The number of groups per species was restricted by the area of the PCPPA. Reproduction beyond the carrying capacity was assumed to be zero.

"Snarability." This parameter was included to specify the probability that an animal is killed if it comes in contact with snares. The assigned parameter values are species-specific. They depend on group size, population size, animal behavior, and visibility to poachers.

"Shootability." This parameter was included to specify the probability that an animal is killed if it is exposed to poachers. This parameter captures the impact of any poaching methods other than snaring (for example, shooting or hand collection). The assigned parameter values are species-specific. They depend on group size, population size, animal behavior, and visibility to poachers.

The poaching activities and the resulting wildlife loss were characterized by the following parameters and relationships:

Poacher groups. Poachers are assumed to operate in groups. This parameter was included in the model to define the number of poacher groups present in the PCPPA at any given time. The model assumed that poacher groups operate independently from each other.

Poacher range. This parameter was included to define the territory targeted by poacher groups. Each group was assigned a territory delineated by a specified number of grid cells, which is the same for all groups. Groups were assumed to move within their territories with an equal probability of occupying any grid cell at any time. One of the grid cells was assumed to be the poacher base. Grid cells were drawn at random from all grid cells located within the PCPPA. The probability of a grid cell being chosen was weighted by its accessibility value. The model assumes that poacher groups do not operate in grid cells below a specified accessibility value. Grid cells below this threshold value were excluded. The grid cell selection protocol can be readily adjusted in case additional data of poacher movements become available. Grid cell selection can be set to be completely random, based on habitat quality values, or based on both habitat quality and accessibility values.

Poaching effort. This parameter was included to define the number of days per month any poaching group operates within their territory. The model assumes that each poacher group sets a snare line with a specified snare density in each grid cell of their assigned poacher range. The model assumes further that snare density is constant across all grid cells targeted by poacher groups. Even though poacher groups are assumed to operate independently, poacher ranges may intersect. If multiple poacher groups target the same grid cells, snare density was assumed to increase proportionally.

Poacher relocation. Poacher groups were assumed to move their base camp and poaching territory from time to time. This was simulated by introducing a poacher relocation parameter that specified the probability that poacher groups move camp and territory. Poacher relocation was allowed at each time step and simulated by drawing a new set of grid cells following the applied selection protocol.

New poachers. New poacher groups were assumed to enter the PCPPA. This was simulated by introducing a parameter that specified the probability that new poacher group enters the PCPPA if a current poacher group was removed by a patrol. The probability was specified such that it simulates the effects of law enforcement delivered by patrols. It was assumed that effective law enforcement reduces the probability of new groups replacing groups that were removed by patrols. The entrance of new poachers was allowed at each time step.

The patrol activities (see Section 2.2) and the resulting poaching reduction were characterized by the following parameters and processes:

Patrol effort. This parameter was included to specify the number of patrol days per month and the number of grid cells patrolled per month. The model specified that the number of grid cells patrolled each day equals the number of grid cells patrolled per month divided by the number of patrol days per month. The model thus allows the investigation of the effect of different levels of patrol intensity (varying the number of cells patrolled per day or month) and patrol frequency (more intense patrolling over a shorter time period versus less intense patrolling over a longer time period).

Patrol range. This parameter was included to define the area targeted by patrols. Each patrol was assigned an area delineated by a specified number of grid cells, which is the same for all patrols. Patrols were assumed to move through their areas with an equal probability of occupying any grid cell at any time. Grid cells were drawn at random from all grid cells located within the PCPPA. The probability of a grid cell to be chosen was weighted by its accessibility value. The model assumes that patrol routes do not include grid cells below a specified accessibility value. Grid cells below this threshold value were excluded. The grid cell selection protocol that defines patrol routes can be readily adjusted in case additional data of poacher movements become available. Grid cell selection can be set to be completely random, based on habitat quality values, or based on both habitat quality and accessibility values.

Patrol relocation. Patrols were assumed to be assigned new patrol routes each month. This was simulated by introducing a patrol relocation parameter that specified the probability that patrols are assigned a new patrol route. Poacher relocation was allowed at each time step and was simulated by drawing a new set of grid cells following the applied selection protocol.

Poacher apprehension. Based on the range of each poacher group and patrol, the model calculated the probability that a poacher group enters a gird cell that contains a patrol on any given day. The probability of apprehension on any given day was calculated by multiplying the probability of a poacher group entering a patrolled grid cell with the specified probability that a patrol apprehends a poacher group if encountered. The probability of apprehension on any given day was converted into a monthly probability value. Using this monthly probability value, whether or not a group was apprehended, was determined stochastically.

Snare removal. Based on the range of each of poacher group and patrol, the model calculated the probability that a patrol enters a gird cell that contains a snare line on any given day. The probability of snare removal on any given day was calculated by multiplying the probability of a patrol entering a snared grid cell with the specified probability that a patrol removed a snare line if found. The probability of snare removal on any given day was converted into a monthly probability value. Using this monthly probability value, whether or not a snare line was removed, was determined stochastically. The model assumed that if a snare line was dismantled, then all snares within that snare line were removed.

Poacher base removal. Based on the range of each of poacher group and patrol, the model calculated the probability that a patrol enters a gird cell that contains a poacher camp on any given day. The probability of camp removal on any given day was calculated by multiplying the probability of a patrol entering a grid cell that contains a camp with the specified probability that a patrol removed a camp if found. The probability of camp removal on any given day was converted into a monthly probability value. Using this monthly probability value, whether or not a camp was removed, was determined stochastically.

Animal deaths were characterized by the following parameters and relationships:

Animal deaths through snaring. Based on the range of each poacher group, the model calculated the probability that an animal group enters a gird cell that contains a snare line with the specified snare density on any given day. The probability of death on any given day was calculated by multiplying the probability of an animal group entering a snared grid cell with the "snarability" value. The probability of death on any given day was converted into a monthly probability value. Using this monthly probability value, whether or not a group was snared, was determined stochastically. A snared group was removed from the simulation. The model assumed that for species that live in families the whole group was snared.

Animal deaths through other poaching method. Based on the range of each of poacher group, the model calculated the probability that an animal group enters a gird cell that contains a poacher group on any given day. The probability of death on any given day was calculated by multiplying the probability of an animal group entering a poached grid cell with the "shootability" value. The probability of death on any given day was converted into a monthly probability value. Using this monthly probability value, whether or not a group was killed was determined stochastically. A killed group was removed from the simulation. The model assumed that for species that live in families the whole group was killed.

In summary, the model simulates the following set of core processes:

1. *Animal maturation and reproduction for each target species.*
2. *Reductions in number of poachers because of patrols.*
3. *Reductions in number of snares because of patrols.*
4. *Possible introduction of new poachers.*
5. *Movements of poachers.*
6. *Movements of patrols.*
7. *Animal deaths from illegal poaching through snaring for each target species.*
8. *Animal deaths from illegal poaching by any other method but snaring for each target species.*

These processes were simulated at a monthly "time step." The model enabled simulations over any number of months.

The patrols may generate additional impacts on wildlife diversity currently not included in the model such as habitat protection through a reduction of illegal logging and illegal collection of nontimber forest products.

Step 3: Data collection:

The model was informed by established knowledge (both empirical and conceptual) and existing data from Lao PDR, South-East Asia, and elsewhere. Data were obtained from a wildlife survey,[4] expert consultations, data banks,[5] and the scientific and "gray" literature. In general, the amount and quality of data available were extremely limited. Due to this limitation, the model was based on a range of assumptions as set out under Step 2. The assumptions made reflect the current understanding of species-specific characteristics as well as the relationships between wildlife populations, poaching activities, and planned patrolling activities within the PCPPA.

The model was designed to cope with a "data poor" context. It is stochastic to account for the uncertainties surrounding the obtained data and established knowledge. Nevertheless, the model predictions are understood as first approximations. The model is readily adjustable once additional data collected by the patrol teams (see Section 2.3) become available. Additional data enable the periodic review and adjustment of the model assumptions and refinement and improvement of overall model validity.

Step 4: Model estimation:

The collected data and established knowledge were used to assign values to each of the model parameters. The model was estimated multiple times ("runs") to account for random errors associated with its stochastic specification.

Step 5: Sensitivity analysis:

A comprehensive sensitivity analysis was conducted to check the model's reliability, validity, and robustness. Based on these checks, the overall model performance was deemed acceptable.

3.2 Estimating production functions

A bioeconomic model can be used to predict the effectiveness of management actions (inputs) in producing environmental service outcomes (outputs). The predicted parameter values associated with PES scheme relevant inputs and outputs form the basis for estimating production functions. A production function may contain multiple input variables but only one output variable. Some production activities do however produce multiple outputs. For instance, growing sheep may produce

---

[4] The wildlife survey was commissioned by the project 'Effective Implementation of Payments for Environmental Services schemes in Lao PDR'.
[5] IUCN List of Threatened Species (https://www.iucnredlist.org/).

wool meat and leather. Similarly, a PES scheme may target the provision of more than one output. In such cases, more than one production function has to be estimated. For example, planting trees in a catchment may yield better water quality downstream as well as providing habitat for endangered species. The same inputs produce multiple outputs.

The first step in estimating a production function is to predict output at varying levels of each input using the bioeconomic model. Output predicted at a zero input level forms the baseline. For example, if no labor is committed to catchment restoration, then the water quality downstream is established as the baseline. Note that this baseline may not be the levels of output currently observed. Water quality may continue to deteriorate without the restoration effort. Alternatively, some regeneration of trees may occur even without the application of labor and this would help to improve water quality. The range of the nonzero input levels needs to match the range of inputs that are potentially provided by suppliers. Additional output generated through the PES scheme is then calculated as the difference between output predicted at a zero input level and output predicted at each of the nonzero input levels. For example, the probability of a storm surge event causing damage to a coastal village may be reduced from 20% to 5% with the planting and development of buffering mangrove forests.

The simulated (or observed) data on output (the dependent variable in the production function) and inputs (the independent variables) enable the estimation of a mathematical function that best fits the data. The so estimated production function specifies the relationship between the quantity of inputs used and the quantity of output produced. The function can be used to calculate the marginal product of each input, that is, the input's productivity. The marginal product of a particular input measures the increase in the quantity of output because of using an additional unit of that input, holding all other inputs constant. If the production function is differentiable, the marginal product of any input quantity is represented by the derivative of the production function at that quantity.

A marginal product is not only a measure of input productivity but can also be interpreted as a "conversion factor" between inputs and outputs: how much output can be expected from an extra unit of input. This is crucial for the design of PES schemes in which the buyers purchase outputs (Chapter 4) but the sellers supply inputs (Chapter 5). Conversion factors enable the linking of inputs with outputs. This link is a necessary (but not sufficient) condition to bringing buyers and sellers together (Chapter 6).

The Lao PES scheme

The bioeconomic model was used to predict the effectiveness of antipoaching patrols (inputs) in protecting wildlife diversity (output). Wildlife diversity is represented by the model parameter "population size of each of the 18 target species." The value of this parameter was predicted for different levels of effort (numbers of patrol days per months) over a 3 year time period. The predicted values were converted into average values per year. Effort levels ranged from no patrols to the maximum total number of patrols possible under the scheme conditions. The parameter value predicted without any patrols was used as the baseline. The model predicted the baseline levels of species diversity to decline over time. Additional output generated through patrolling was calculated by taking the difference between the baseline value and the values estimated for each input level. Another model parameter of interest was the reduction of the number of snares. The predicted values were suited for use as an estimate of the cost of bonus payments for snare removal.

The model parameter "population size of each of the 18 target species" was disaggregated and converted to generate two output variables: "species present in the PCPPA" (an indicator for species diversity) and "percentage reduction of animals poached" (an indicator for population sizes). A separate production function was estimated for each of the two outputs. Padé approximations were used for production function curve fitting.[6]

Both production functions exhibit diminishing marginal productivity as expected: The patrol effectiveness decreases with an increasing number of patrols (see Fig. 3.1). The production functions were differentiated to calculate the respective marginal products (output generated per additional patrol).[7] Due to data limitations, potential negative effects on wildlife populations associated with high patrol densities (such as negative effects caused by disturbance to wildlife through the presence of more people) could not be simulated. Marginal products could drop to zero or become negative if such effects were included in the model.

The marginal product of each output was used to convert outputs into inputs: that is, how many units of input are required to produce an extra unit of output. This enabled linking demand estimated in "output space" (Chapter 4) with supply estimated in "input space" (Chapter 5). This link facilitated the estimation of "market" prices by matching demand with supply (Chapter 6).

3.3 Challenges and limitations

Developing bioeconomic models in a "data poor" context presents major challenges. These challenges can be addressed, at least to some extent, by choosing a stochastic model specification, relying on established knowledge, transferring data from neighboring countries, and using expert opinions[8] to inform the model. However, it is important to acknowledge that initial model predictions have to be interpreted as first approximations. A periodical model assessment and adjustment process is crucial to increasing model validity over time.

---

[6] Spider (Python 3.5) and Excel Solver (Excel, 2013) were used for curve fitting.
[7] SageMath 7.6 was used to differentiate the production functions.
[8] The expert survey used to inform the development of the model is presented in annex 1.

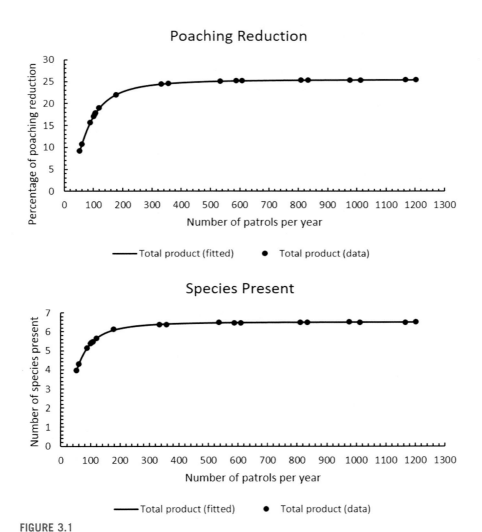

FIGURE 3.1

Total production of anti-poaching patrolling.

Source: Scheufele G., Bennett J. & Kyophilavong P. (2018) 'Pricing biodiversity protection: Payments for Environmental Services schemes in Lao PDR', Land Use Policy, Vol 75, pp.284–291.

The development of a model that is sufficiently complex to represent reality requires the use of a modeling program or language. This may present a challenge in contexts where the necessary skills are limited among the people responsible for the development of a PES scheme. Model development, assessment, and adjustment may be severely restricted. This challenge may be addresses to some degree by including training as a core element in the development and implementation of a

PES scheme. Using software that can be used free of charge supports any training effort in this regard.

The main purpose of a model is to understand better a complex system by reducing its complexity. Every model embodies trade-offs between realism, precision, and generality. Increasing a model's degree of generality decreases its realism and precision. That is, models that are developed to simulate a large number of environmental management actions (inputs) and environmental service outcomes (outputs) might be of limited use in facilitating the estimation of the associated benefits and costs. Consequently, developing a single model for PES schemes that aim to increase the provision of a large number of environmental services through the supply of a large number of environmental management actions may not be feasible.

References and further readings

Special acknowledgement is given to Michael Renton, Marit Kragt, Eric Hay and Chanthavy Vongkhamheng for developing the bioeconomic model.

Publications of team members and collaborators of the Lao PES project:

Hay, E., Kragt, M., Renten, M., Vongkhamheng, C., 2017. Research Report 11: Modelling the Effects of Anti-poaching Patrols on Wildlife Diversity in the Phou Chomvoy Provincial Protected Area. Crawford School of Public Policy, The Australian National University, Canberra.

Scheufele G., Bennett J., Kyophilavong P. (2018) 'Pricing biodiversity protection: Payments for Environmental Services schemes in Lao PDR', Land Use Policy, Vol 75, pp. 284−291.

Scheufele, G., Vongkhamheng, C., Kyophilavong, P., Tsechalicha, X., Bennett, J., Burton, M., 2016. Research Report 9: Providing Incentives for Biodiversity Protection: Anti-poaching Patrolling in the Phou Chomvoy Provincial Protected Area. Crawford School of Public Policy, Australian National University, Canberra.

Vongkhamheng, C., 2015. Research Report 8: Phou Chomvoy Provincial Protected Area: A Biodiversity Baseline Assessment. Crawford School of Public Policy, Australian National University, Canberra.

Publications of authors external to the project:

Conteh, A., Gavin, M.C., Solomon, J., 2015. Quantifying illegal hunting: a novel application of the quantitative randomised response technique. Biological Conservation 189, 16−23.

Frank, R., 2003. Microeconomics and Behavior. McGraw-Hill, Boston.

Hansel, T., Moore, C., O'Kelly, H., Vongkhamheng, C., Eickhoff, G., Ferrand, J., 2013. National Protected Area and Wildlife (NPAW) Project Report. The Wildlife Conservation Society, Vientiane.

IUCN (International Union for Conservation of Nature). Red-List of Threatened Species. http://www.iucnredlist.org/. Cited 1 June 2016.

MacKenzie, D., Nichols, J., Lachman, G., Droege, S., Royle, J., Langtimm, C., 2002. Estimating site occupancy rates when detection probabilities are less than one. Ecology 83, 2248−2255.

MacKenzie, D., Nichols, J., Royle, J., Pollack, K., Bailey, L., Hines, J., 2005. Occupancy Estimation and Modeling: Inferring Patterns and Dynamics of Species Occurrence. Academic Press, San Diego, California.

Mankiw, N.G., 1998. Principles of Economics. The Dryden Press, Harcourt Brace College Publishers, Fort Worth.

Royle, J.A., Dorazio, R.M., 2008. Hierarchical Modeling and Inference in Ecology. Academic Press/Elsevier, New York.

Schoen, R., 2010. Dynamic Population Models. Springer, Netherlands.

Watson, F., Becker, M.S., McRobb, R., Kanyembo, B., 2013. Spatial patterns of wire-snare poaching: implications for community conservation in buffer zones around National Parks. Biologiacal Conservation 168, 1–9.

Demand—what buyers want

4

Buyers express their interest in environmental services by their willingness to pay. The more people are willing to pay for an environmental service, the stronger are their preferences for that service. In this manner, the value that people enjoy from consuming more of an environmental service is reflected by their willingness to pay. In other words, willingness to pay is a measure of the "marginal benefits" enjoyed from consuming more. People's willingness to pay is not only determined by the strength of their preferences. People are also constrained in expressing their preferences through their ability to pay. Hence, people's income and their wealth, as well as their preferences to use their available resources to purchase other goods and services also drive their willingness to pay.

Preferences are also, in part, determined by the amount of the service people already have. The general rule is that people are willing to pay less for more of a service, the more they already have. This is one of the better-known "laws" of economics: as the quantity consumed increases, the marginal benefit enjoyed decreases. First year economics students are often given the chocolate cake example of this law: The first piece of chocolate cake enjoyed for morning tea is enjoyed greatly but the additional value enjoyed from the fifth slice is no doubt much smaller and probably close to zero. That means people are willing to pay a lot more for the first slice than the fifth. In the context of environmental services, people are less likely to be willing to pay very much for an extra hectare of protected forest ecosystem if there were already a million hectares of that ecosystem set aside than if there were only 10 ha given protected status.

These drivers of willingness to pay—and preferences—are commonly described by what economists know as a demand function. This is the relationship between the amount people are willing to pay for an extra unit of a good or service and the amount of that good or service already available. This is a negatively sloped function: more available means less willingness to pay. Other driving factors, like disposable income, shift this function: for example, more income means more willingness to pay.

This chapter details the processes that can be used to understand Payments for Environmental Services (PES) scheme buyer preferences through the estimation of the demand function for environmental services.

Buying and Selling the Environment. https://doi.org/10.1016/B978-0-12-816696-3.00004-1

4.1 Estimating demand

One of the key characteristics of environmental services, and a major source of impetus for PES style interventions, is that they are "open access" (see Box 1). The important outcome of this characteristic is that potential buyers are unlikely to reveal the strength of their preferences through conventional markets. The "free-rider" motive is likely to ensure that people fail to reveal the full extent (if any) of their preferences through market exchange. Hence, the equally conventional method used by economists to quantify the extent of demand—interrogating market data—cannot be used in the case of open-access environmental services.

Recognizing this deficiency, economists have developed a range of "nonmarket" demand valuation techniques. Essentially what these techniques do is to allow the estimation of willingness to pay for differing levels of environmental services. Some use an "ex post" approach that involves the collection of data relating to the experiences of potential buyers. Others are "ex ante" in their approach. This means that data relating to potential future scenarios are collected and analyzed. Some of these techniques rely on observations of people's behavior in markets for goods and services that are somehow related to the nonmarketed environmental service in question. These are known as "revealed preference" techniques. Others, known as "stated preference" techniques, involve potential buyers being asked about their preferences in hypothetical scenarios.

In the next section, a range of these nonmarket valuation techniques is presented as a "tool box" for PES scheme designers who find it necessary to inform the PES "pseudomarket" with data on buyer preferences.

4.2 Nonmarket valuation techniques

4.2.1 Revealed preference techniques

Where nonmarketed environmental services are consumed alongside marketed goods and services, the relationship of "complementarity" can be exploited to estimate environmental services' values. The prerequisites for the use of revealed preference techniques extend beyond that of complementarity. They also require data to be available regarding the use of the associated marketed good or service. These data are usually only available for environmental services that have "use values" and where experiences are relevant.

The two most commonly applied revealed preference techniques are the travel cost method (TCM) and the hedonic pricing technique (HPT).

The TCM relies on the complementary relationship existing between travel and recreational experience. Data regarding the expenses incurred by people traveling to enjoy environmental services can be analyzed to infer values enjoyed from the nonmarketed environmental services. The data analysis requires the estimation of a relationship between the frequency of visitation to a site of interest over a period of time and the costs of traveling to that site. This "trip generation function" is then used to simulate the number of site visitations under different scenarios of hypothetical entrance fees. The assumption made is that visitors would respond to an entrance fee in the same manner as they respond to their existing travel costs: the higher the costs the lower the visitation rate. The simulations of visitation and hypothetical entrance fee enable the estimation of a standard demand curve for the recreational experience (an environmental service) and from that, the conventional willingness to pay measure of value per visit.

Data to implement a TCM are usually collected at the site of the recreational experience. Visitors are asked questions regarding their trip to the site, with particular attention being given to the amount spent on their travels and how often they visit.

The HPT also uses a complementary relationship but between the consumption of a nonmarketed environmental service and a marketed good that has the environmental service as a characteristic. The most frequent context for applications is when an environmental service has an influence on the price of real estate. For instance, the amenity provided by proximity to or a view of green open space in an urban setting may influence the willingness to pay of potential buyers of properties. Implementing the technique requires the estimation of a relationship between prices paid for properties and the set of variables that have an impact on those prices, one of which is the environmental service of interest. The "implicit price" of the environmental service is the coefficient of that amenity in the equation estimating the price relationship. It is important to note that the implicit price is the contribution to property prices made by the environmental service. Strictly speaking, this is not a measure of the buyers' well-being that results from the characteristic. It is a part of the cost paid by the buyer, rather than a benefit enjoyed. To estimate the appropriate well-being measure, it is necessary to move to a second stage of analysis that relates buyer characteristics (such as income and age) to the property characteristics and the price paid.

The data requirements of an HPT application are heavy. Large volumes of property sales and characteristics data are required to estimate accurately the impacts of all factors that impact on price. To move to the second stage of analysis, buyer data are also required. Property data may be accessible from government records of sales but frequently on-site inspections are needed to measure the environmental services attribute that is otherwise not available. Buyer data are rarely available, and this factor has limited the application of the HPT to the first stage in most cases.

These two revealed preference techniques have the advantage of being based on observed market data. This grounds the estimates achieved in the reliability of actual preferences as demonstrated through buyer behavior. However, they both have limitations. Both rely on past data relating to use. They are therefore of little use when "nonuse" values are relevant and where new contexts that do not relate well to experience are under consideration. Furthermore, there are specific issues relating to each method. For instance, problems occur for the TCM when a single trip involves multiple destinations and when the costs of travel time are incorporated. For the HPT, the implicit price estimates derived from the first stage of analysis are not conceptually consistent with the economic measures of buyer well-being.

4.2.2 Stated preference techniques

Partly in response to these shortcomings, stated preference techniques for nonmarket valuation have been developed. These techniques involve surveying potential buyers to find out how they would react to a range of new but hypothetical contexts that involve changes in the availability of the environmental service of interest. Survey respondents are thus asked to "state" the strength of the preferences in a questionnaire rather than "reveal" them in an associated market as occurs in the revealed preference techniques.

The ability to invoke hypothetical future scenarios means that stated preference techniques are able to cover the full spectrum of environmental services—use and nonuse, present and future. Furthermore, they are capable of producing estimates of nonmarket value that are consistent with economic principles. However, because of the hypothetical nature of the questions asked of survey respondents, there is the danger of potential buyers failing to state their strength of their preferences accurately—either deliberately in an attempt to "game" the process or inadvertently because of a lack of pertinence. Because of this danger, the design and implementation of stated preference surveys is of particular importance.

The two most prominent stated preference techniques are the contingent valuation method (CVM) and choice modeling (CM).

The CVM, in its earliest form, simply involved survey respondents being asked what they would be willing to pay for more environmental services. The incentive that this style of questioning provided to respondents to bias their answers led to the development of the "dichotomous choice" version of the CVM. This involves subsamples of respondents being asked if they were willing to pay a prespecified amount for a defined improvement in environmental service provision. Analysis of the proportion of the subsamples that agree to pay the differing amounts provides estimates of average willingness to pay. A limitation of the CVM in this format is the cost of application. Only one hypothetical scenario can be presented to each respondent and statistically representative subsamples must be interviewed for each of the willingness to pay bid values that are used. This means that large numbers of respondents are required to elicit values for only one scenario. Where there are multiple possible scenarios under consideration, the costs become infeasible for most projects.

In contrast, CM is able to produce estimates across multiple scenarios given just one application. The concept underpinning CM is that environmental services are attributes of future resource use options. A respondent to a CM questionnaire is asked to make choices between alternative futures that are described by these attributes taking on different levels. An empirical analysis of the choices people make to allow for the estimation of the average willingness to pay of respondents for units of each attribute. With these estimates, composite values of different future combinations of environmental services at differing levels can be derived. Hence, CM not only accesses the flexibility advantages of CVM being a stated preference technique, but it also provides cost savings.

Similar to CVM, CM is also subject to concerns regarding estimate reliability. Questionnaires for both CVM and CM applications must adhere to certain design rules to mitigate against estimate inaccuracies. Most importantly, the context of the questionnaire must be realistic so that respondents understand that their answers are relevant to a real PES or policy-making situation. There must be a belief in respondents that their answers will lead to consequences, both in terms of environmental service outcomes but also in them having to face some financial consequences.

4.3 Selecting a nonmarket valuation technique

The choice between demand estimation techniques is highly dependent on the context of the PES scheme under development. Different PES schemes will involve different environmental services with different extents of market interaction. For some PES schemes, it may even be possible to rely on markets to deliver accurate information on the extent of demand for an environmental service. This would require that the environmental service is not "open access." Where nonmarket estimation techniques are required, revealed preference techniques are preferred where they can be applied. They offer the advantage of being based on actual choices (and hence trade-offs) made by people. However, their application relies on the environmental services involved having some "use" component to their value and a history of provision. These conditions are frequently absent in PES scheme contexts. Then, stated preference techniques will be required and due care needed to ensure that the hypothetical choices they involve are not biased.

For some contexts, multiple environmental services will be involved. The possibility is that different elements within the "bundle" of environmental services will require the use of different demand estimation tools. This will involve first the careful definition of the environmental services involved (set up under the "scope" of the services) and then, where necessary, the separation of the elements for the valuation task. This is particularly problematic if there are "overlaps" across the types of environmental services involved and the different groups that are involved as buyers. The prospect of "double counting" must be avoided. With this in mind, multiple demand estimation techniques can be employed and their results aggregated. Stated preference techniques have the advantage of being able to encompass multiple types of environmental services (for instance, use and nonuse values) in the one application. However, different groups of people enjoying different types of environmental services will usually require separate demand estimation studies, even if they use the same technique (applied in different ways).

Where direct markets are absent, all environmental service demand estimation methods require the collection and analysis of primary data. This implies transaction costs for a PES scheme, which in turn reduces the prospects of a PES scheme being viable in terms of delivering net improvements in peoples' well-being. In some circumstances, these transaction costs can be significantly reduced using the "benefit transfer" method. Benefit transfer is not a nonmarket valuation technique per se. Rather it involves the use of values estimated in contexts that are similar to those characterizing the current case. Hence, "source values" are used to estimate "target values."

Although the use of benefit transfer saves time and money and may make a PES scheme viable, it comes with caveats. Most importantly, source studies must be available. Given that the number of nonmarket environmental valuation studies remains limited due to the comparative infancy of the discipline, this is a considerable constraint. Even where values of the specific environmental service have been estimated in other studies, care must be exercised in the use of benefit transfer. Value estimates are highly context specific: Variations in contexts will cause variations in benefits. Hence, contexts must align relatively closely. That means consistency between biophysical settings, similarity of environmental management actions and environmental services outcomes as well as homogeneity across the people whose values are being estimated. Furthermore, the validity of the source study estimates must be held with confidence and adjustments made to take account of differences in currencies and the impacts of inflation over time. International databases such as the Environmental Valuation Reference Inventory (evri.ca) are useful to implement benefit transfer.

4.4 Choice modeling

The development of the stated preference technique, choice modeling, has represented a major advance in environmental nonmarket valuation because of its superior flexibility to cover a range of benefit types and to deliver estimates of multiple benefit scenarios in a single application. Its appeal for application in the design and implementation of PES schemes in primary cases or as a source method for benefit transfer is clear.

As outlined in Section 4.4, the choice modeling technique involves respondents to a questionnaire making choices between alternative future scenarios of environmental service management, for example, PES scheme outcomes. The flexibility afforded by CM is derived from its use of attributes or characteristics to describe the alternative futures. Different futures involve the attributes taking different levels. The choices made between the alternatives by survey respondents show their willingness to make trade-offs between the attributes. Sufficient variability between the choice alternatives to allow quantitative modeling is achieved using experimental designs to assign attribute levels to each alternative. With one choice alternative always being the "do nothing" option, and with one attribute being a monetary cost of making a change, a "willingness-to-pay" value can be estimated for improvements in each of the nonmonetary attribute. Because individual attributes are valued, any number of possible combinations of attribute levels making up different alternative PES outcomes can be valued.

Because some of the CM attributes can relate to use values and others to nonuse values, the full spectrum of values can be estimated separately or together. Furthermore, the future alternatives presented to respondents are not restricted to existing or past options. Novel circumstances can be investigated.

A potential weakness of choice modeling is its reliance on the presentation of hypothetical future scenarios to respondents. Of particular importance in this regard is the hypothetical nature of the cost to be borne by respondents if a change is to occur, for example, if a PES scheme is introduced. To address the prospect of biased values being estimated because of the hypothetical context, a range of strategies has been developed by CM practitioners. For example, the "payment vehicle" used to collect the money charged for change has to be as realistic as possible and respondents must understand and believe that their answers to the choice questions will be "consequential," that is, there will be a change to environmental conditions and they will have to pay for them as a result of their answers to the survey. Hence, CM questionnaires must come with a sense of authority and purpose. Part of this is ensuring respondents understand that payment is not voluntary but instead will be forced on respondents. Voluntary payment requests run the risk of respondents attempting to "free ride" by refusing to pay in the hope that others will pay sufficient to see the scheme progress. This is the case even when the environmental services involved are "open access." Information provided in a CM questionnaire about the context and alternatives must be accurate but clear and concise. The choices themselves must be readily understandable so that people can associate their answers with the PES scheme being considered along with any associated public policy measures.

The use of alternative payment vehicles in CM questionnaires has an added advantage for PES scheme design. It enables the testing of alternative ways of raising funds for the scheme. If the consequentiality of the questionnaire is successfully transmitted to respondents, their reactions to different funding mechanisms can be judged.

The design of a choice modeling questionnaire needs to provide information to respondents on the environmental service for which they are going to be asked to pay. The mechanism for the service to be provided (the process of supply) and the consequences of them not buying it are also fundamental components of the questionnaire. The choices provided to respondents relate to alternative outcomes of environmental service provision (including one that involves no additional services being purchased). Attributes used to describe the outcomes may be related to the types of benefits provided by the environmental service (for example, use and nonuse values). However, the complexity of the information to be disseminated through the questionnaire poses some challenges. The effective communication of that information requires the use of language and presentation that is appropriate to the respondents. Differing levels of literacy across respondent groups will for instance require different styles of presentation, including the use of graphics and illustrations, as well as differing levels of language complexity. Where different languages are spoken by the respondent group, particular care is required. The oral presentation of the questionnaire may be problematic with respondents feeling more comfortable with reading text.

Bias is a concern for questionnaire design. Not only does the questionnaire need to be considered consequential by respondents to avoid hypothetical bias, it needs to be perceived by respondents as presenting the information (and the case to purchase environmental services) in an unbiased manner. Where the questionnaire is presented to respondents in a face-to-face interview, there is also the danger that the interviewer can induce biased answers. This might come about because of a desire of the respondent to please the interviewer, or in contrast, the respondent may become frustrated by an interviewer and bias their answers as a consequence. The possibility of interviewer bias calls for the developments of strategies to neutralize the impetus. These involve the thorough training of interviewers but also the incorporation of questionnaire design features such as graphics, interspersing oral delivery with show cards and the removal of interviewer influence in the respondents' choice option selection process.

The choice sets presented to respondents offer alternative environmental service provision, with each option described by attributes that take on different levels. Selecting the "mixture" of attribute levels that constitutes an alternative and how the alternatives are mixed into the choice sets requires the use of an "experimental design." This is because the number of alternatives that is possible is simply too large for any respondent to handle. A sample of the "full factorial" of combinations that could make up alternatives needs to be drawn and an experimental design performs that sampling role. The simplest of experimental designs involves taking a random sample of alternatives from across all those that are possible. Such an experimental design requires a relatively large number of respondents to answer the choice sets to collect sufficient data to estimate statistically sound results. This process can be made more efficient if the experimental design can use some existing data on respondents' preferences. If some "priors" relating to the relative strength of preferences across attributes are available, these can be used in the experimental design process to increase efficiency. This enables a smaller sample of respondents to be used or it allows for greater statistical significance in the results.

There is a range of different means of collecting choice modeling data. These relate to the differing ways in which questionnaires can be delivered to respondents. Personal interviews have the advantage of engaging with respondents and hence achieving a strong response rate. They are however relatively expensive and come with the prospect of interviewer bias, as discussed previously. An increasingly popular alternative is delivery through the internet. Web-based surveying is comparatively cheap and quick but can face difficulties of achieving a representative sample when the penetration of computers and computer literacy is limited. Attention to the material presented over the web by respondents is also a potential concern. Web-based surveying has become increasingly widespread given the increasing weakness of response rates coming from mail delivered surveys. They are also relatively slow and expensive given the need to print and post extensive amounts of material. Phone sampling is problematic for choice modeling questionnaires given that the material to be presented to respondents is necessarily complex.

The way a sample of respondents is drawn for a choice modeling application is dependent on the mode of delivery. Where personal interviews of residents are being used, sampling can be based on records of the population being sampled (such as an electoral roll) or geographically. For the latter, areas from which the sample is being drawn can be subdivided into grid cells on a map and respondents drawn from randomly selected cells. Otherwise, tourist respondents for example may be sampled on site using a random interception protocol. Web-based samples are usually achieved through accessing market research companies' "panels" of email addresses where people have previously agreed to be available for participation. Concerns regarding individual privacy in many advanced economies have restricted access to population databases from which samples can be drawn. In these cases, and in less developed economies where information regarding populations is also restricted but because of it not being collected, geographic sampling and web-based protocols are likely to be more feasible.

Once the choice data have been collected from respondents, they are analyzed using logit models in which the probability of choosing an alternative is explained by the levels that the attributes take in each alternative as well as a range of socio-economic characteristics of each respondent. The coefficients on each of the attributes indicate the impact of the attribute on respondents' choices. The ratio of an attribute coefficient against the coefficient of the payment vehicle attribute is a measure of the willingness of respondents, on average, to sacrifice some of the payment vehicle (money) to gain more of the environmental service attribute. This is the willingness to pay for a single unit increase in the provision of the environmental service described by the attribute. It is therefore the monetary measure of the respondents' value for the environmental service.

Choice models can account for the heterogeneity of respondents' preferences by including socioeconomic characteristics in the model specification and using parameter estimates that are distribution rather than point estimates.

The Lao PES scheme

Buyers. The CM method was used to estimate the willingness to pay for the provision of the specified environmental services (see Section 2.1) *of the residents of the urban districts of Vientiane City ("residents") and international tourists visiting the Lao People's Democratic Republic (PDR) ("tourists") (see Section 2.5).*

Survey mode. The survey was conducted through face-to-face interviews. The interviewers were students from the Faculty of Economics and Business Management at the National University of Laos.[1]

---

[1] Intensive training for the student interviewers as well as survey supervision and management was provided by the ACIAR project "Effective Implementation of Payments for Environmental Services in Lao PDR."

Survey material. The survey material included a survey protocol for the interviewers, an information card to establish the interviewer's credentials, a questionnaire script that included instructions for the interviewer, show cards, a choice booklet, and an answer sheet.[2] The tourist questionnaire is provided in Annex 2.

The survey protocol specified general instructions, the sampling protocol, and the code of practice. For example, the interviewers were instructed to select respondents by strictly using the sampling strategy, not to record the name of the respondents to ensure confidentiality and anonymity, to follow the questionnaire instructions, and to use the show cards. The code of conduct specified, for example, that interviewers must follow exactly the wording of the questionnaire script without adding or omitting information, never attempting to give an answer to the question being asked on behalf of the respondent or to draw conclusions "hypothesizing" about what the answer may be, and never making any promises to the respondents.

The show cards illustrated the area of the PCPPA, a typical village of the region around the PCPPA, examples of target wildlife species, and an example choice question. Some of the show cards contained background information, information on the attributes, and the more complex or sensitive questions. The show cards were used in combination with the questionnaire script. As the pilot survey revealed that the reading skills of the tourists were superior to their oral skills (given the range of first languages spoken), the use of show cards reduced communication barriers between the Lao interviewers and the (foreign) tourist respondents. The show card also improved communication in cases where the slow speaking pace of the interviewers seemed to frustrate the respondents. Communications were facilitated further by the extensive use of symbols and images in the printed material.

The choice booklets contained the choice questions. Respondents were requested to record their choices in the booklet (without being observed by the interviewer) and to submit them in a sealed envelope. This approach was used to guarantee confidentiality and to minimize any response bias associated with a desire to please the interviewer or a reluctance to reveal true preferences.

The questionnaire script (and some of the show cards) had to be tailored to the two buyer groups. This resulted in differences with respect to the filter questions, questions on sociodemographics, and the payment vehicle (see later). The questionnaire used for the resident respondents was delivered in Lao. The tourist respondents were interviewed in English. Limited language skills of the interviewers prevented the delivery of the tourist questionnaire in any language other than English. Respondents without sufficient English language skills could not be interviewed.

Interviews. Each respondent interview was structured in four sections. It was started with a range of filter questions to support the stratified sampling process. Tourist respondents were asked about their purpose of travel, visa requirements,

---

[2] The survey material and survey logistics were tested and revised through focus groups comprised of residents and a pilot survey of tourists. Practical obstacles prevented focus groups being conducted for the tourists and a pilot survey of residents being performed.

and country of origin. Resident respondents were asked if they were citizens or permanent residents of the Lao PDR. In the second section, respondents were given information on the PCPPA, on the current status of and threat to wildlife diversity in
that area, and on options of how the PCPPA could be managed in the future to protect wildlife diversity more effectively. The third section was dedicated to the choice
questions. Each respondent was requested to make a sequence of five choices between three management options: one option not involving any new management actions at no additional cost and two options involving new management actions at an
additional cost. Each option was described in terms of attributes (see later). An
example choice question was used to inform the respondents about the outcomes
of choosing each of these three options (see Fig. 4.1).

　　Choice attributes. The choice options (and therefore the choice outcomes) were
described by five attributes that set out the environmental and social services provided and the associated costs of purchasing these services (see Fig. 4.2). Three noncost attributes associated with the specified environmental services KNOWLEDGE
("species diversity" and "poaching reduction") and WATCHING ("tourist access")
were specified. An additional noncost attribute was included to describe the choice

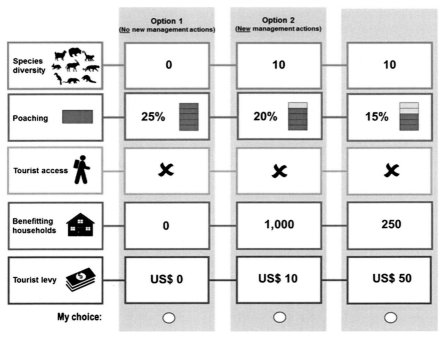

FIGURE 4.1

Choice set example (tourist split-sample).

Source: Reprinted with permission from Scheufele G. & Bennett J. (2019). Valuing Biodiversity Protection:
Payment for Environmental Services Scheme in Lao PDR. Environment and Development Economics, Cambridge
University Press ©2019.

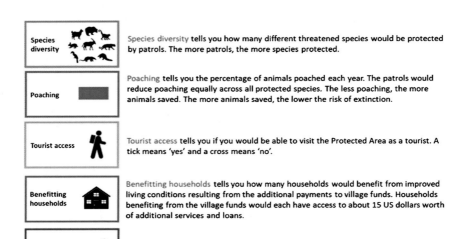

FIGURE 4.2

Attribute description example (tourist split-sample).

Source: Reprinted with permission from Scheufele G. & Bennett J. (2019). Valuing Biodiversity Protection: Payment for Environmental Services Scheme in Lao PDR. Environment and Development Economics, Cambridge University Press ©2019.

options. This attribute—"benefitting households"—described the social service "knowledge of improved living conditions of the people living in close proximity to the PCPPA" (henceforth-called HOUSEHOLDS). The respondents were told that the people's living conditions would be improved through the PES scheme through payments to the village development funds, which would allow the provision of additional public services and access to loans. The payment vehicle associated with the cost attribute had to be tailored to each of the two buyer groups. Tourist respondents were presented with a one-off tourist levy ("tourist levy") while resident respondents faced a monthly household payment collected through the electricity bill ("household payment"). The levels of the cost attribute had to be customized to the different levels of disposable income/wealth and purchasing power of the two buyer groups. The levels of all other attributes were identical across the two buyer groups. All attributes were described by symbols to address communication problems with illiterate resident respondents and tourist respondents with limited English language capabilities. The levels of all attributes and their respective coding are shown in Table 4.1.

Experimental design. Bayesian S-efficient experimental designs (500 Halton draws) were generated to construct the choice sets presented to the tourist and resident respondents. Ngene 1.1.2 was used to calculate a separate design for each of the two buyer groups. The information on the priors used to calculate the designs was collected during the pilot survey of tourists and the focus groups conducted

Table 4.1 Attribute levels and coding.

Attribute	Attribute levels
Tourist levy—tourist split-sample only (one-off tourist levy—US$)[a]	$0
	$5
	$10
	$20
	$50
Household payment—inhabitant split-sample only (monthly household payment through electricity bill—Lao ₭)[a]	₭ 0
	₭ 5000
	₭ 10,000
	₭ 20,000
	₭ 40,000
Species diversity (number of species)	0
	5
	10
	15
	20
Poaching (percentage of animals poached per year)	25%
	20%
	15%
	10%
	5%
Tourist access (availability of tourist access to protected area)	1 (yes)
	−1 (no)
Benefitting households (number of households located in close proximity to the protected area that would benefit from improved living conditions as a result of an additional payment to village funds)	0
	250
	500
	1000
	1500

[a] *$US1 = ₭8177.68 (27.01.2017 Oanda.com).*

Source: Reprinted with permission from Scheufele G. & Bennett J. (2019). *Valuing Biodiversity Protection: Payment for Environmental Services Scheme in Lao PDR. Environment and Development Economics*, Cambridge University Press ©2019.

with residents. Each experimental design comprised 20 choice questions that were grouped into four blocks of five choice questions. The order in which the choice questions of each block were presented to the respondents was randomized to reduce possible ordering effects. This approach produced 20 different choice booklets per buyer group, which were assigned to the respondents at random.

Data collection. A random sample of the population associated with each buyer group was drawn to collect the data required to estimate the choice models. Data collection was conducted from December 5, 2015 to December 15, 2015. Tourist respondents were sampled in the departure lounge at Wattay International Airport in

Vientiane City. The interviewers applied a random sampling method based on the seating locations within the departure lounge. The data collection timetable covered the departure times of all international flights. The resident interviews were conducted at the respondents' homes. Maps delineating district boundaries were unavailable, which prevented stratified sampling. As for the tourist survey, a random sampling approach was applied. The interviewers used the Google earth app on their mobile phones to locate the starting points for random household sampling that were randomly selected by the survey supervisor a priori.

Econometric modeling. Mixed logit models were used to analyze the collected choice data. The cost parameter was specified as nonrandom. All noncost parameters were specified as random assuming a normal distribution (1000 Halton draws) to capture potential preference heterogeneity across respondents. The model included a generic error component to account for potential differences in error term variances between management options (and thereby relaxing the IID assumption) and was specified as a panel model to accommodate repeated choice (each respondent made five choices). The Newton–Raphson algorithm was applied to facilitate the estimation process. The models for both buyer groups were estimated in STATA 13. The Krinsky–Robb parametric bootstrapping procedure (10,000 repetitions) was used in the estimation of the implicit prices to account for sampling errors.

Response rate. The response rates of the tourist and the resident split-samples (excluding protest respondents) are 60% (sample of 354 respondents) and 42% (sample of 206 respondents), respectively. Tourist respondents without English language capabilities were classified as nonresponses. Protest respondents were defined as respondents who were not willing to reveal their preferences. Their preferences remained unknown. The split-samples contain 9 and 23 protest respondents, respectively.

Sample representativeness. The two split-samples were characterized by a range of sociodemographics to check if they were representative of their corresponding "buyer group" populations (see Tables 4.2 and 4.3). Population-level data were sourced from the Lao Census of Population and Housing (2015) and on the Statistical Tourism in Laos (2015) report. Sample representativeness was checked through a sequence of Chi-Square tests. Both split-samples were found to be statistically different from their respective buyer groups with respect to the sociodemographics for which population-level data were available. The tourist respondents, on average, visited the Lao PDR for a longer time period than the tourist buyer population. Asian tourists were underrepresented, while Australians/New Zealanders, Europeans, and North Americans were overrepresented. This difference may have been caused, to some degree, by the limited English language capabilities of some of the Asian respondents. The resident respondents, on average, were better educated and younger than their corresponding population. The resident split-sample contained more females than the group population. Furthermore, government employees, state enterprise employees, and unpaid family workers were overrepresented. These differences may have been caused by the interview schedule. The interviews had

Table 4.2 Socio-demographics of tourist split-sample.

Variable		PCPPA	Population
Gender	Female	40.78%	NA
	Male	59.42%	NA
Highest level of education	Primary education	0.00%	NA
	Secondary education	11.59%	NA
	Tertiary education	88.41%	NA
Age	18–24	21.45%	NA
	25–29	30.43%	NA
	30–39	18.55%	NA
	40–49	14.78%	NA
	50–59	8.70%	NA
	60–69	4.64%	NA
	70–79	1.16%	NA
	80 and older	0.29%	NA
Country	Africa and Middle East	2.03%	1.66%
	Americas	19.77%	13.90%
	Asia and pacific	26.16%	50.51%
	(Australia and New Zealand)	11.63%	6.16%
	Europe	52.03%	33.93%
Average household income		$81,813	NA
Average size of travel party		2.00 people	NA
Average stay		8.3 days	7.5 days
N (respondents)		354	

Source: Reprinted with permission from Scheufele G. & Bennett J. (2019). Valuing Biodiversity Protection: Payment for Environmental Services Scheme in Lao PDR. Environment and Development Economics, Cambridge University Press ©2019.

to be carried out between 8:30 am and 6:00 pm. Within the Lao context, sampling outside this time frame was presumed to be unsafe for the interviewers and perceived as rude by the respondents.

Preferences. The results of the econometric analysis showed that the parameter estimates associated with the "tourist levy"/"household payment," "poaching reduction," "tourist access," and "benefitting households" were statistically different from zero (at least at the 5% confidence level). This suggests, holding everything else constant, that the respondents preferred lower costs to higher costs, lower poaching levels to higher poaching levels, having the opportunity to access the area to not having that opportunity, and more households benefiting from improved living conditions to less households. The estimate associated with "species diversity" was only statistically different from zero (at the 1% confidence level) in the tourist split-sample. This suggests that tourist respondents preferred more species diversity to less, whereas resident respondents were indifferent toward that attribute. The econometric analysis revealed that the standard deviations of the random parameters were

Table 4.3 Socio-demographics of resident split-sample.

Variable		PCPPA (%)	Population (%)
Gender	Female	58.74	50.46
	Male	41.26	49.54
Education	No education	1.94	1.31
	Less than 6 years primary education	7.28	23.52
	Primary education	8.74	1.10
	Secondary education	29.13	49.91
	Vocational education	19.90	11.63
	Tertiary education	33.01	12.53
Age	18–24	10.68	21.54
	25–29	12.14	15.95
	30–39	22.82	22.63
	40–49	25.73	15.88
	50–59	17.48	11.40
	60–69	7.28	6.29
	70–79	3.40	3.71
	80 and older	0.49	2.60
Main activity last 12 month	Government employee	26.70	13.71
	Private employee	2.43	18.36
	State enterprise employee	12.62	2.61
	Employer	3.88	1.14
	Self-employed	4.37	17.15
	Unpaid family worker	20.87	3.02
	International organization or NGO	0.49	0.56
	Unemployed	3.40	2.79
	Student	3.40	21.11
	Household duties	14.56	14.24
	Other	7.28	5.31
Average household income		37,883,100 ₭	NA
Average household size		5.4	4.6
Household head	Yes	46.12	
	No	53.88	
N (respondents)		206	

Source: Reprinted with permission from Scheufele G. & Bennett J. (2019). Valuing Biodiversity Protection: Payment for Environmental Services Scheme in Lao PDR. Environment and Development Economics, Cambridge University Press ©2019.

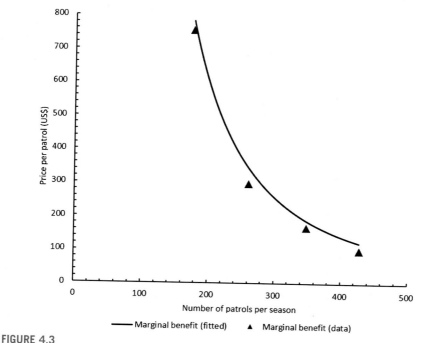

FIGURE 4.3

Demand function ('busy' season).

Source: Scheufele G., Bennett J. & Kyophilavong P. (2018) 'Pricing biodiversity protection: Payments for Environmental Services schemes in Lao PDR', Land Use Policy, Vol 75, pp. 284-291.

4.6 Challenges and limitations

Estimating demand through nonmarket valuation techniques poses some challenges. One of these challenges is caused by time frame differences between demand and supply. Demand of environmental services is typically perpetual, whereas supply is based on finite supplier contracts. Finding payment vehicles that enable temporal harmonization can be challenging.

Another challenge is associated with environmental services for which willingness to pay for an additional unit of output varies across output quantities. That is, the willingness-to-pay estimates are different for each level of input provided by the sellers. The presence of so-called nonconstant willingness-to-pay estimates adds to the complexity of their conversion from "output space" into "input space."

Survey-based nonmarket valuation presents a range of challenges, especially in a developing country context. For example, sampling may be challenging if spatial information on statistical divisions and population-level data are not available or sociocultural norms limit the sampling schedule. Such sampling barriers reduce the likelihood of drawing a sample that is representative of the population. As a result, the extrapolation from sample to population may be biased. If census-based population data are not available, sample-based data may have to be used for extrapolation. Without census-based data, sample representativeness cannot be checked. In those circumstances, it remains unknown whether or not the extrapolated demand estimate is biased.

Customizing the survey material and the sampling protocols to a developing country context may introduce additional challenges associated with sociocultural norms, limited language skills of interviewers, as well as low literacy levels and language diversity of respondents. Furthermore, respondents may not trust the assurances of the interviewers that they remain anonymous and their answers are treated confidential. The application of procedures that ensure anonymity and confidentiality are crucial to encourage the respondents to reveal their true willingness to pay.

The survey material and sampling protocols may also have to be customized if a PES scheme involves multiple buyer groups. The more split-samples that are required to cover the multiple buyer groups, the higher are the overall transaction costs of the scheme development.

It must also be acknowledged that the use of stated preference nonmarket valuation techniques raises the prospect of estimates that are biased due to respondents not reporting their preferences accurately. This can occur because respondents fail to give the necessary attention to their answers because the contexts of the questioning are hypothetical. It may also occur if respondents seek to bias their answers in an attempt to manipulate the survey outcome and hence the design of the PES scheme to their advantage. The issue of biased estimates from stated preference methods has been the subject of a great deal of research effort. This effort has resulted in method designs that counter incentives for biased results. However, practitioners must be mindful of the potential for bias and take appropriate measures to deal with those incentives. The potential for bias varies from application context to application context and developing country contexts are particularly challenging in this regard. These challenges relate to the varying "buyer groups" (locals and foreigners) having differing incentives and the prospects of being able to develop payment vehicles that deliver the necessary "consequentiality" of the survey in the minds of respondents.

References and further readings

Special acknowledgement is given to Phouphet Kyophilavong for supporting the data collection effort.

Publications of team members and collaborators of the Lao PES project:

Scheufele, G., Bennett, J., 2019. Valuing Biodiversity Protection: A Payments for Environmental Services Scheme in Lao PDR. Environment and Development Economics. Published online May 2019.

Scheufele, G., Bennett, J., 2017. Can payments for ecosystem services schemes mimic markets? Ecosystem Services 23, 30—37.

Scheufele, G., Bennett, J., 2017. Research Report 13: Valuing Biodiversity Protection: Payments for Environmental Services Schemes in Lao PDR. Crawford School of Public Policy, Australian National University, Canberra.

Scheufele, G., Bennett, J., 2013. Research Report 1: Payments for Environmental Services: Concepts and Applications. Crawford School of Public Policy, The Australian National University, Canberra.

Scheufele, G., Bennett, J., Kragt, M., Renten, M., 2014. Research Report 3: Development of a 'virtual' PES Scheme for the Nam Ngum River Basin. Crawford School of Public Policy, The Australian National University, Canberra.

Publications of authors external to the project

Alberini, A., Kahn, J.R., 2006. Handbook on Contingent Valuation. Edward Elgar Publishing, Cheltenham.

Barbier, E., Hanley, N., 2009. Pricing Nature: Cost-Benefit Analysis and Environmental Policy. Edward Elgar Publishing, Cheltenham.

Bennett, J., 2011. The International Handbook on Non-market Environmental Valuation. Edward Elgar Publishing, Cheltenham.

Bennett, J., Birol, E., 2010. Choice Experiments in Developing Countries. Edward Elgar Publishing, Cheltenham.

Bennett, J., Blamey, R., 2001. The Choice Modelling Approach to Environmental Valuation. Edward Elgar Publishing, Cheltenham.

Bliemer, M.C.J., Rose, J.M., 2005. Efficiency and Sample Size Requirements for Stated Choice Studies. Report ITLS-WP-05-08. Institute of Transport and Logistics Studies, University of Sydney.

Ferini, S., Scarpa, R., 2007. Designs with a-priori information for non-market valuation with choice experiments: a Monte Carlo study. Journal of Environmental Economics and Management Science 53, 342—363.

Frank, R., 2003. Microeconomics and Behavior. McGraw-Hill, Boston.

Government of Lao PDR, 2016. Results of the Lao Population and Housing Census 2015. Vientiane.

Johnston, R.J., Rolfe, J., Rosenberger, R.S., Brouwer, R., 2015. Benefit Transfer of Environmental and Resource Values. Springer, Dordrecht.

Krinsky, I., Robb, A.L., 1986. On approximating the statistical properties of elasticities. The Review of Economics and Statistics 68, 715—719.

Lancaster, K., 1966. A new approach to consumer theory. Journal of Political Economics 74, 217—231.

Lew, D.K., Wallmo, K., 2017. Temporal stability of stated preferences for endangered species protection from choice experiments. Ecological Economics 131, 87—97.

Louviere, J., Hensher, D., 1982. On the design and analysis of simulated choice or allocation experiments in travel choice modelling. Transportation Research Record 890, 11—17.

Louviere, J., Hensher, D., Swait, J., 2000. Stated Choice Methods. Analysis and Application. Cambridge University Press, Cambridge.

Louviere, J., Woodworth, G., 1983. Design and analysis of simulated consumer choice experiments or allocation experiments: an approach based on aggregate data. Journal of Marketing Research 20, 350—367.

Mankiw, N.G., 1998. Principles of Economics. The Dryden Press, Harcourt Brace College Publishers, Fort Worth.

McFadden, D., 1974. Conditional logit analysis of qualitative choice behaviour. In: Zarembka, P. (Ed.), Frontiers in Econometrics. Academic Press, New York.

McFadden, D., 1980. Econometric models for probabilistic choice among products. Journal of Business 53, 13–29.

Revelt, D., Train, K., 1998. Mixed logit with repeated choices: households' choices of appliance efficiency level. The Review of Economics and Statistics 80, 647–657.

Sándor, Z., Wedel, M., 2001. Designing conjoint choice experiments using managers' prior beliefs. Journal of Marketing Research 38, 430–444.

Tietenberg, T., Lewis, L., 2009. Environmental and Natural Resource Economics. Pearson International Edition, Boston.

Tourism Development Department, 2016. 2015 Statistical Tourism in Laos: Ministry of Information, Culture and Tourism. Government of Lao PDR, Vientiane.

Train, K., 1998. Recreation demand models with taste differences over people. Land Economics 74, 230–240.

Ward, F.A., Beal, D., 2000. Valuing Nature with Travel Cost Models. Edward Elgar Publishing, Cheltenham.

Supply—what sellers want

5

Sellers express their interest in supplying an environmental service by the payment they are willing to accept in return. Their "willingness to accept" depends on their opportunity costs of supply. The higher their opportunity costs, the higher their willingness to accept. The opportunity costs are expected to differ among sellers. Yet, as a general rule for all sellers, the opportunity costs depend on the extent of service supply. The more sellers supply, the higher are their opportunity costs. For example, sellers are more likely to be willing to accept a small payment for supplying the first rather than the hundredth extra hectare of a protected forest ecosystem. This is a reflection of the relative scarcity of the forest and what else could be done with the forest. The first hectare of forest protected represents a small sacrifice in terms of other opportunities. However, with more and more forest protected, the value of the forest as a producer of timber gets higher and higher because timber becomes more and more scarce.

These drivers of willingness to accept—and therefore their marginal opportunity cost of supply—are commonly described by what economists know as a supply function. This is the relationship between the amount sellers are willing to accept for an extra unit of a good or service and the amount of that good or service already supplied. This is a positively sloped function: more supply means higher willingness to accept. A supply function thus provides information on marginal costs of production over a range of supply quantities.

A conservation auction can be used to reveal the extent of the marginal costs (supply) for environmental services that are not usually bought and sold in markets. It can be designed to estimate sellers' individual marginal cost functions, which—once aggregated across sellers—represent a supply function at the "market" level. This chapter details how to reveal Payments for Environmental Services (PES) scheme sellers' willingness to accept through the estimation of a "market" supply function for environmental services. It begins by detailing the importance of a PES scheme engaging with the community of potential suppliers. This process sets the foundations for the supply estimation process. The chapter then progresses to detail the ways in which conservation auctions can be conducted to gather marginal cost data that are suited to the process of matching with the demand analysis detailed in the previous chapter. They also form the basis of contractual agreements between the PES scheme operators and suppliers.

Buying and Selling the Environment. https://doi.org/10.1016/B978-0-12-816696-3.00005-3

5.1 Engaging the community

Engaging the broader community in the PES scheme development and seller recruitment process is essential to build trust, obtain consent, and secure support. Otherwise, the social cohesion of the community and sustainability of the scheme may be jeopardized.

Successful community engagement begins with the collection of information on the socioeconomic and ecological context. Effective tools for collecting this information may include literature reviews, community surveys, as well as the consultation of local authorities, experts, public servants, and members of the private sector. The collected information can be used to construct community resource profiles. Presenting and discussing these profiles at an early stage during the community consultations can stimulate interest among and encourage engagement of community members.

Community consultations are a key element in the development of a PES scheme. They ensure that community engagement is based on prior, informed, and free consent. Community consultations enable the distribution of information and provide opportunities for questions, comments, suggestions, and concerns to be raised. Community consultations need to be conducted in a culturally acceptable format. For example, providing opportunities for anonymous feedback may increase the number of people who are willing to contribute. Alternatively, separate discussion groups for men and women may inspire women to participate and voice their opinions more forthright than they would otherwise. Information material such as brochures, presentations, photographs, satellite images, or illustrations may increase the effectiveness of community consultations. It is essential that the material is customized to the local context to avoid social exclusion, especially the exclusion of vulnerable people who would be affected by the scheme. For example, language or literacy barriers can be reduced if the material minimizes text and relies largely on illustrations and photographs.

The community consultations may also be used to establish a common understanding and interpretation of statutory legislation and customary laws and regulations relevant to environmental service supply. To achieve this, the broker's current understanding could be presented as an initial input in the consultation process. The communities would then be invited and encouraged to provide feedback, comments, and corrections. This process may support environmental service supply in two ways. First, the collected information may form the legal basis for environmental management actions taken by the sellers. Second, it might reduce environmentally harmful behavior of communities that is solely based in ignorance. The local population may violate legislation merely because they are not aware of their existence or do not understand its meaning in their daily lives.

The community consultations may also provide the setting to conduct conservation auctions and the prerequisite bidding training.

Community consultations are finalized by negotiating the terms and conditions of environmental service supply. The negotiated terms and conditions need to be formalized through legally binding contracts and agreements (see Chapter 7). Otherwise, supply additionality and payment conditionality are unlikely to be achieved.

The Lao PES scheme

The community engagement process was initiated by informal meetings with village leaders. The meetings were used to provide information about the proposed PES scheme and its purpose. During these meetings, the village leaders signaled a general interest in learning more about the scheme. These initial meetings were followed by a community survey involving all eight villages. The purpose of the community survey was to collect more detailed data on the socioeconomic and environmental context of the villages and the Phou Chomvoy Provincial Protected Area (PCPPA). The survey material contained a survey protocol, a questionnaire, satellite images of the villages and the PCPPA, images of wildlife species, and livelihood activity cards. A random sample of 332 households (about a third of each village) was drawn from the village registers. Participation in the survey was voluntary and the names of the respondents were not recorded to ensure anonymity and confidentiality. The interviewers were trained to adhere to the survey protocol and apply core research values including scientific rigor, transparency, and research ethics. The collected data were extrapolated to the household population of all eight villages. Unfortunately, whether or not the sample was representative could not be checked. The relevant population-level data were not available. The survey results were used to construct village resource profiles. The following section presents the highlights of the resource profile averaged across all eight villages.

Village resource profiles:

Households carry out a broad range of activities to secure their livelihoods (see Fig. 5.1). Almost all households perform agricultural activities, whereas only few households are engaged in government and private sector employment. The villages produce crops (96% of households), grow vegetables and fruits (72%), farm livestock (92%), collect nontimber forest products (61%), harvest wildlife (36%), produce handicraft (32%), engage in hired labor (38%), are employed by the government (21%), are employed by the private sector (8%), and perform other livelihood activities (13%).

The percentage of cash earning households across livelihood activities is shown in Fig. 5.2. The most households secure cash earnings through livestock farming (68%), followed by cropping (55%), hired labor (38%), and fruit and vegetable production (31%). Only a few households secure cash earnings through private employment.

The sources of cash earnings are shown in Fig. 5.3. Livestock and crop sales are the source of almost half of village earnings (23% and 25%, respectively). Another substantial source is government employment (19%). Private employment (9%), hired labor (6%), and fruit and vegetable sales (6%) are less important. The least

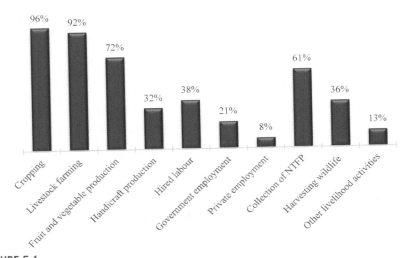

FIGURE 5.1

Percentage of households by livelihood activities.

Source: Household survey conducted within the project 'Effective Implementation of Payments for Environmental Services Schemes in Lao PDR'.

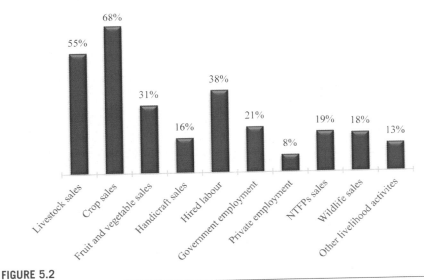

FIGURE 5.2

Percentage of cash earning households by livelihood activity.

Source: Household survey conducted within the project 'Effective Implementation of Payments for Environmental Services Schemes in Lao PDR'.

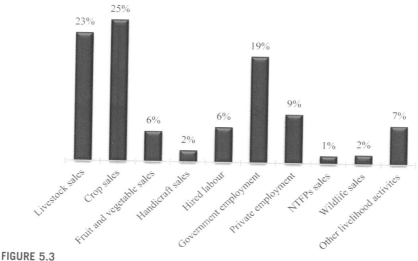

FIGURE 5.3

Percentage of cash earnings by livelihood activity.

Source: Household survey conducted within the project 'Effective Implementation of Payments for Environmental Services Schemes in Lao PDR'.

amount of cash is earned through nontimber forest products and wildlife sales (2% and 1%, respectively).

The relative cash earnings for each livelihood activities are shown in Fig. 5.4. *Private sector employment is by far the most lucrative livelihood activity. The relative earnings secured through private sector employment are almost 14 times higher than that earned through the collection of nontimber forest products. The second most lucrative livelihood activity is government employment. The relative earnings are almost 12 times higher than those secured through collecting nontimber forest products.*

The overall relative cash earnings differ across villages (see Fig. 5.5*). The richest village earns more than three times the poorest village.*

Fig. 5.6 *shows that about one-third of households harvest wildlife. Households hunt or catch wildlife exclusively for subsistence use (15%), exclusively for sale (1%), and for both subsistence use and sale (20%). Households mainly catch and hunt wildlife species they are legally allowed to harvest for subsistence use as long as they comply with the stipulated restrictions. Wildlife is harvested using bows, guns, hunting dogs, hand collection, and snares. Snaring is the most common hunting method: all but 3 of the 18 species included in the household survey are harvested by snares or a mixture of snares and other methods. Snares are particularly problematic as they kill indiscriminately a wide range of ground dwelling animals.*

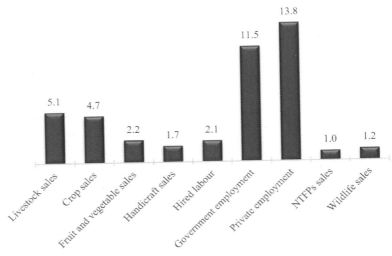

FIGURE 5.4

Relative cash earnings by livelihood activitiy.

Source: Household survey conducted within the project 'Effective Implementation of Payments for Environmental Services Schemes in Lao PDR'.

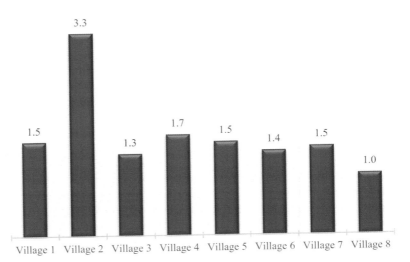

FIGURE 5.5

Relative cash earnings by village.

Source: Household survey conducted within the project 'Effective Implementation of Payments for Environmental Services Schemes in Lao PDR'.

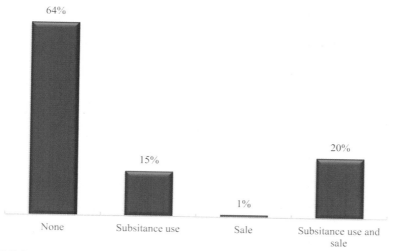

FIGURE 5.6

Percentage of households engaged in wildlife harvesting by purpose.

Source: Household survey conducted within the project 'Effective Implementation of Payments for Environmental Services Schemes in Lao PDR'.

Fish and farm meat are consumed more frequently (about four times per week) than wildlife (about one time per week). This corresponds to the preferences households hold for different types of protein sources. The most preferred source of protein is meat produced through animal rearing (42% of households), followed by fish (37%) and wildlife (21%).

About a third of households (35%) have experienced problems with wildlife. The problems reported include crop raids (30%), attacks on livestock (5%), and attack on people (2%). Nevertheless, most households (85%) think wildlife protection is very important or important (see Fig. 5.7). Only some households think it is only moderately important (9%), of little importance (5%), or unimportant (1%).

The resource profiles suggests that agriculture is the predominant livelihood activity. Alternative employment opportunities are limited and not available to most households. More than half of village cash earnings are secured through agriculture. Yet, the relative cash earnings secured from this livelihood activity are rather low compared to those obtainable in the public or private sector. Public and private sector employment opportunities are, however, only available to a small segment of households. Differences in cash earnings are not just observed across households but also across villages. The cash earnings of the richest village are more than three times higher than those of the poorest village. This indicates that the opportunity costs of time are likely to differ across households within and across villages. The resource profile suggests further that hunting and catching wildlife is crucial for about a third of households to secure their subsistence. They target mainly those

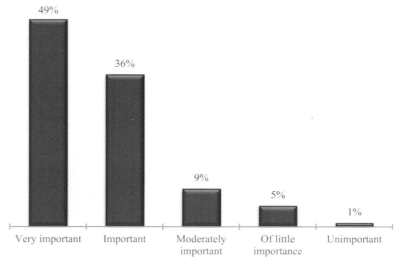

FIGURE 5.7

Percentage of households by view on the importance of wildlife protection.

Source: Household survey conducted within the project 'Effective Implementation of Payments for Environmental Services Schemes in Lao PDR'.

species they are allowed to use for subsistence under Lao legislation as long as they comply with the stipulated restrictions. Even though the consumption of wildlife is an important of their diet, most households prefer farmed meat and fish over wildlife. Limited employment and cash earning opportunities together with the households' hunting behavior, food preferences, and general concern for wildlife protection suggested community engagement and support for the antipoaching patrol scheme to be substantial.

Community consultations:

For community consultations to be effective in building trust, obtaining consent and securing support, they need to be well prepared. This includes the selection and training of facilitators. Under the Lao scheme, GoL officials and villagers were engaged as facilitators to assist in the community consultations. The village facilitators were selected based on the following criteria:

1. *Good understanding of the village situation and a good relationship with villagers.*
2. *Interest in facilitating the community consultations at the village level.*
3. *Ability to work with village groups and confidence to speak and manage the audience.*
4. *Ability to read and write in Lao.*
5. *Willingness to attend facilitation training.*

The selection process accounted for different ethnic minorities and was gender sensitive. The number of village facilitators employed per village was proportional to its population size.

The GoL officials and the selected villagers were trained in two separate sessions to stimulate open and critical discussions. The feedback from the training sessions was used to revise the PES scheme design and the consultation material.

The training of the facilitators was followed by a stakeholder meeting. The meeting was chaired by the Vice Governor of the Bolikhamxay Province and attended by staff of the GoL authorities of Khamkeut and Xaychamphone districts, delegates from the eight target villages, and the trained consultation facilitators. The objectives of the meeting were to present and discuss the overall ideas of the PES scheme, collect feedback, and obtain consent to initiate the community consultations. The feedback from the stakeholder meeting was used to revise the PES scheme design and the consultation material.

The stakeholder meeting was followed by the village-level community consultations. A special effort was made to create an atmosphere that encouraged the villagers to contribute toward the development of the PES scheme and provide honest feedback without the fear of repercussions. The consultations were customized to the local context of each community. This ensured that each community was engaged an inclusive and culturally appropriately manner. Care was taken to stimulate the participation all villagers independent of their social status, gender, and ethnic group. For example, both women and men participated in the consultations and contributed in the development of the scheme but the most remote village that has had only limited exposure to outside influences.

The community consultations were conducted in two rounds. The conservation auctions and the prerequisite bidding training were conducted directly after the second round. Each round of consultations as well as the bidding training were organized in segments along a range of topics. The project developed consultation and conservation auction training manuals for facilitators. The conservation auction training manual is provided in Annex 4. *Each segment was introduced by an oral presentation that provided an overview of the topic as well as more detailed information. The presentation was supported mainly by illustrations, maps, satellite images, and photographs to avoid the exclusion of illiterate villagers from the consultations. Numerous examples and exercises were used to explain more complex topics and keep the villagers interested and engaged. A booklet communicating the most important information and most of the material presented during the presentation was developed by the project and distributed among the villagers. The booklet is presented in* Annex 3. *The booklet facilitated the discussions among the villagers during and after the community consultations. After the presentation of each topic, the villagers were invited to engage in a discussion and encouraged to ask questions, provide comments, make suggestions, and voice concerns. The discussions were either conducted in the full plenum or small subgroups. It was made clear that neither the names of villagers nor the full discussions would be recorded. Anonymous summaries of each subgroup were presented to the full plenum by the facilitators. At the conclusion of the consultations, all villagers had a final opportunity to ask questions, provide comments, make suggestions, and voice concerns regarding anything they regarded relevant.*

Villagers who preferred to withhold feedback during the consultations had the opportunity to engage in writing via *anonymous feedback forms. The villagers were instructed not to provide their names and to submit the feedback form within a sealed envelope. The envelopes were collected after a few days to encourage discussions in the absence of the facilitators.*

The first round was organized along 11 topics:

1. *PES scheme concept*

This segment was used to explain what a PES scheme is and how it works.

2. *Opportunities of participation*

This segment was used to outline that both the communities as a whole and individual villagers would be given the opportunity to participate in the PES scheme.

3. *Antipoaching patrol scheme design*

This segment was used to present and discuss what the patrol scheme would look like and how it would work.

4. *Community benefits*

This segment was used to present and discuss the benefits the village communities and patrol teams would receive.

5. *Conservation auctions*

This segment was used to explain how conservation auctions work and how the patrol payments would be determined.

6. *Payment transfer*

This segment was used to explain payment transfer from the brokers to the communities. Special attention was given to explaining the advantages of a cashless transfer mechanism and the functioning of bank accounts.

7. *Penalty system*

This segment was used to present and discuss the penalty system in case community commitments would not be honored.

8. *Mechanism for grievance, conflict resolution, and redress*

This segment was used to present and discuss how a community as a whole and individual villagers could file a complaint, resolve conflicts, and secure redress.

9. *Contracting*

This segment was used to present and discuss how the commitments of the community and the patrol teams would be formalized through community conservation agreements and patrol contracts, respectively.

10. *Invitation to register an expression of interest*

   This segment was used to explain how a village as a whole entity and individual villagers could register an expression of interest in participating in the PES scheme. Village interest had to be registered through an "expression of interest form" signed by the village authorities. Villagers interested in the patrolling scheme needed to form groups of five villagers, of which two or three had to be members of the village militia and at least two had to be literate. Group registration was only valid if the "expression of interest form" was signed by all five group members. Village groups and villages had the opportunity to register their interest up to 2 weeks following the completion of the first round of the community consultations. This deadline allowed communities to discuss internally in the absence of the project staff and facilitators. It was made clear that an expression of interest was not an obligation to participate in the patrolling scheme but a signal of interest to participate in the bidding for patrolling and the corresponding training workshop.

   Special attention was given to the most remote village. Given its unique strategic location relative to the PCPPA, engaging several patrol teams from this village was considered essential. However, only one team from this village registered their interest. The facilitators suggested that, due to a lack of senior provincial government representatives, the consultation process did not reassure the villagers enough that the PES scheme had the full support of the GoL. An additional consultation session with a more senior representative of Provincial Agricultural and Forestry Office (PAFO) present was held to encourage more villagers to engage in the patrol scheme. As a result, two more teams registered their interest. Given that an additional consultation session was required and the remote location of the village, the bidding training, and the auctions was conducted by village facilitators without the assistance of the project team.

11. *Process of community engagement*

   This segment was used to set out the next steps of the community consultations. The objective of this segment was to provide assurances that process would continue.

   About 50 feedback forms were submitted. Given that some of the feedback was not articulated during the consultations indicates that the anonymous and confidential submission procedure was successful. Generally, the feedback suggested that the villagers and the village authorities perceived the scheme as a feasible and effective approach of protecting wildlife diversity and improving their livelihoods for current and future generations.

   The feedback focused on the following issues:

- *Payment amounts*
- *Payment transfer mechanism*
- *Payment schedule*
- *Insurance cover*
- *Restrictions on the use of wildlife*
- *Penalties for noncompliance*
- *Seasonal differences with respect to opportunity costs of time*

All feedbacks received during and after the consultations were reviewed and used to revise the PES scheme design.

The first round of community consultations triggered a strong interest in PES scheme participation. On behalf of their respective communities, the village authorities of all eight villages submitted an expression of interest. Additionally, a total of 65 village teams expressed their interest in engaging in the antipoaching patrol scheme. The number of team submissions differed across villages, ranging from 1 to 14 teams per village. Even though 65 teams submitted an expression of interest, only 40 of these teams provided forms signed by all members and/or complied with the conditions on team size and composition. After follow-up assistance, 55 valid team submissions were obtained.

Before the second round of village-level community consultations, the facilitators received further training and were encouraged to provide feedback on the drafts of the community action plan and the community conservation agreement that was further developed in collaboration with the village communities during the first round of consultations. The feedback was used to revise the PES scheme design and the consultation material (including the draft of the community action plan and the community conservation agreement).[1]

The second round of community consultations was organized along seven topics:

1. *Recap of suggested PES scheme*

This segment was used to recall the information provided in the first round of the consultations as a basis for further inputs and discussions.

2. *Response to feedback received during the first round of community consultations*

This segment was used to present all the feedback that was received from the communities. It was explained which elements of the scheme were changed as a result of the feedback and which were kept unchanged and why. The aim of this segment was to build trust in and ownership for the scheme.

3. *Discussion of the community resource profiles*

This segment was used to present and discuss the community resource profiles. The objective of this segment was to familiarize the communities with the wildlife species present in their vicinity, raise awareness of current wildlife use, and spark interest in wildlife diversity protection.

4. *Legal restrictions on wildlife use*

This segment was used to explain the legal restrictions on wildlife use. The objective was to familiarize the communities with these restrictions and raise awareness of (unknowingly) committed legal violations.

---

[1] *The templates of the signed versions of the community action plan and the community conservation agreement are presented in* Annexes 5 and 6.

5. *Development of community action plans*

This segment was used to develop community action plans. This collaborative process was initiated by presenting and discussing draft versions that were developed on the basis of the results of the first round of consultations. The community action plans contain five core actions:

Action 1: Protection of wildlife diversity in the PCPPA (see Section 2.2)

Action 2: Establishment of a VDF

Each village community adopts a set of principles developed by LuxDev to manage and audit their VDF: According to this principle, the VDF can be used as a grant scheme and a credit scheme. It was explained that it would be the decision of each community to decide if they wanted to have a grant scheme only or a mixture of both. The credit scheme would be a village-owned microfinance system from which members would be able to take loans. However, as poor and vulnerable households may be unable or reluctant to take loans, at least 20% percentage of the payments to the fund would have to be available for the grant activities that benefit poor and vulnerable households whose livelihoods might be negatively affected through the wildlife protection actions. The grant scheme would be managed by the Village Development Committee. The objective of the grant scheme would be to assist the community in making small-scale, one-time investments in activities that improve the living standards and quality of life of the population. Men and women from all households would be involved by democratic voting to select activities. Activities that are expected to cause substantial environmental negative impacts would be excluded from funding. Each community would support the training of VDF managers that would enable them to perform their management and auditing tasks in accordance with the VDF principles.

Action 3: Establishment of a mechanism for grievance, conflict resolution, and redress at the village level (as set out in the Community Engagement Framework developed under the Protected Area and Wildlife Project (PAWP)).

Action 4: Support of a participatory process to monitor and evaluate the social impacts of the PES scheme (as set out in the Community Engagement Framework developed under the PAWP).

Action 5: Implementation of physical and cultural resources "chance-find" procedures to mitigate against damage or loss through antipoaching patrolling (as set out in the Community Engagement Framework developed under the PAWP).

6. *Development of community conservation agreements*

This segment was used to develop community conservation agreements that formalized the community action plans. This collaborative process was initiated by presenting and discussing draft versions that were developed on the basis of the results of the first round of consultations.

7. *Next steps*

This segment was used to set out the next steps of the development and implementation of the PES scheme. The objective of this segment was to provide assurances that the process would continue.

The second round of community consultations followed the same procedures as the first round. The community feedback that was collected during and after the consultations was used to revise the community action plans and community conservation agreements. The feedback received during the second round was mostly concerned with the definition and clarification of terms rather than with content. Care was exercised that revisions were consistent across and approved by all eight villages. The community action plans and community conservation agreements were signed by all eight villages. This result suggests that the consultations were effective in engaging the broader community in the development of the PES scheme and the associated community action plans and community conservation agreements.

Bidding training:

The second round of the community consultations was also used to conduct the conservation auctions and the prerequisite training. The training was organized along five topics:

1. *Recap of suggested antipoaching patrol scheme*

This segment was used to recall the information provided in the first round of the consultations. The objective was to ensure that all bidding teams understood all elements of the patrol scheme: patrol scheme design, patrol tasks, patrol team management, and patrol equipment.

2. *Presentation of Environmental Code of Conduct*

This segment was used to explain the details of the Environmental Code of Conduct each patrol team would have to comply with. It was made clear that patrol teams are expected to be a role model for the wider community and that any violation of the statutory legalization on wildlife protection by any member of the patrol team during antipoaching patrolling or off-duty would be treated as a violation of this Environmental Code of Conduct.

3. *Physical and cultural resources chance-find procedures*

This segment was used to explain the procedures to mitigate against damage or loss of physical and cultural resources through antipoaching patrolling (as set out in the Community Engagement Framework developed under the PAWP).

4. *Patrol contracts*

This segment was used to review the patrol contract template[2] and explain its content in great detail. It was crucial that the bidding teams understood all contract elements (see Section 7.2).

5. *Bidding process*

This segment was used to familiarize villagers with bidding process and procedures of conservation auctions. The training was designed to include mock auctions (involving goods and services other than patrolling) that provided the teams with the opportunity to gain bidding experience through "learning by doing." The goods and services that were used in the mock auctions were familiar to the villagers, such as purchase orders for berberine vine and casual labor in construction. A fundamental aspect of the training was to explain the concept of opportunity costs as a basis for deciding the number of patrols they would be willing to perform given the presented range of prices. The concept of opportunity costs was explained by comparing a daily wage of casual work and the payment offered per patrol. The villagers were taught to factor into their bids all the costs incurred by performing patrols, including those associated with participating in the conservation auction, transportation to and from the PCPPA.

Villagers were also trained in team bidding (as opposed to individual bidding) and to submit a separate bid for each season. The bidding forms used for the patrol bidding was designed in the same way as the forms used in the mock auctions. Immediately before the bidding, facilitators explained the details of the patrol auction form and reviewed the definitions of the terms (such as "patrol," "number of patrols," "team price per patrol") used in the form. Throughout the training process, the facilitators encouraged the villagers to ask any questions they might have had. The facilitators also assisted the teams during the bidding but were instructed not to fill in the forms on behalf of the teams or to provide teams with cost information or patrol numbers.

Even though the concept of bidding was a novelty to most villagers, the training was successful in communicating novel and complex concepts such as the concept of opportunity costs of time. Yet, the time and effort required varied across villages. These differences may be explained by differences in education levels and exposure to market.

5.2 Estimating supply

The absence of conventional markets for open-access environmental services prevents potential sellers from revealing their marginal costs of supply through market transactions. Market data to quantify the extent of supply are therefore not available.

The marginal costs of supply are only known to the sellers. They are "hidden" from the PES scheme designer. A conservation auction is a tool that can be used

---

[2] The template of the final version of the patrol contract is provided in Annex 7.

to estimate these costs. This is achieved by having potential suppliers revealing their respective "willingness to accept" for different levels of environment service supply in a competitive bidding process.

In cases where sellers supply environmental management actions rather than environmental service outcomes, some of the inputs used to produce the outputs might be bought and sold in conventional markets. Information on the marginal costs of supplying these inputs might be sourced from market data.

5.3 Conservation auctions

A conservation auction is a widely used tool to generate information on open-access environmental service supply. In an ordinary auction, buyers bid a price to purchase a good or service from sellers. A prominent example is a real estate auction. In a reverse auction (also called procurement auction), sellers bid to provide a good or service to buyers in return for a payment. A conservation auction (also called conservation tender) is a reverse auctions applied in the context of environmental service supply. Bidders tender the amount of a service they are willing to supply at differing prices.

A conservation auction can be used to estimate the costs of supplying either an environmental service (output) or inputs used to produce an output (see Chapter 3). In an output-based conservation auction, prospective sellers bid competitively to supply a specified quantity of an environmental service in return for a payment. The bids reflect the amount bidders are willing to accept—and thus their marginal costs—for supplying an output. For example, prospective sellers may bid to refrain from clearing a protected forest ecosystem. Estimating the marginal costs of setting aside an additional hectare might be straightforward. Sellers calculate the forgone profits they could have earned if they cleared the forest and sold the timber.

However, the production function underlying the supply of an output may be unknown to the sellers. The sellers do not have sufficient information to estimate their opportunity costs of output supply. In this case, a conservation auction can be designed to estimate the costs of supplying the required inputs to produce the output. For example, an input-based conservation auction may be conducted to estimate the costs of the inputs associated with planting trees and installing sediment barriers at the riverbank to produce improved water quality. Or they may refrain from clearing coastal mangrove forests to decrease the impacts of high tides or the frequency of storm surges. Naturally, the use of an input-based conservation auction presumes the capacity of the broker to generate information on the production function. This information is needed to convert input quantities into output quantities. Otherwise, the supply estimated in "input space" cannot be matched with the demand estimated in "output space."

The conservation auction design described here generates individual marginal cost functions of sellers of output (or input). In the bidding process, each seller offers the quantity of output (or input) they are willing to supply within a specified time frame over a range of prices specified by the broker. The quantity of output (or input) offered is thus a function of price. The sequence of price–quantity pairs generated for each bidder represents their individual marginal cost functions. Any indirect returns enjoyed by the sellers may be included in their bids. Sellers may account for these returns such that they may offset their supply cost to some degree.

Although all bidders are presented with the same price range, output (or input) quantities offered at each price are likely to differ across bidders. The observed differences may be explained by differences in opportunity costs of supply across sellers. For example, some sellers might have a range of high-income employment opportunities, which are unavailable to others. Selecting a price range that is effective in capturing supply cost differences across bidders is crucial. The endpoints of an effective price range may represent the "best guesses" of the lowest and highest opportunity costs likely to be experienced by the sellers. Determining the number of price levels within the determined range involves a trade-off between the costs of the cognitive burden placed on the bidders and the costs of the errors associated with fitting marginal cost functions to the auction data. A large number of price levels reduce the fitting errors but increase the cognitive burden and vice versa.

All bidders need to understand fully the terms and conditions of environmental service supply, including the specification of output (or input) and the payment regime. The output (or input) sellers bid to supply needs to be specified rigorously. This involves specifying the unit of measurement as well as the quality standards each unit of output (or input) must meet. Specifying the desired output (or input) in quantitative and qualitative terms is a necessary condition to enable the monitoring and enforcement of supply (see Chapter 7). Supply monitoring and enforcement, in turn, are essential to ensure that the supply is additional (see Section 1.5).

The rigorous specification of output (or input) provides the basis for a performance-based payment regime, which is a necessary condition to ensure the payments made to the sellers are conditional on supply. Sellers are paid per unit of output (or input). The payments that successful bidders receive in return for supplying the specified quantity of output (or input) can be financial or in-kind in nature. However, in-kind payments (for example food) need to be convertible into a common unit (usually money). Otherwise, information on supply cannot be matched with demand data. It also needs to be noted that in-kind payments may increase the transaction costs borne by the broker. Alternatively, this additional cost may be internalized by reducing the payments and thus lowering the profits enjoyed by the sellers. All bidders also need to understand fully the payment schedule (the timing and frequency of the payments) and the payment transfer (the mechanism by which the payment are transferred from the broker to the seller).

Effective conservation auctions are governed by a set of specified rules and procedures. These rules and procedures must be understood and obeyed by both the bidders and the broker. Achieving the required level of understanding may involve intensive training, which adds to the overall transaction costs of the PES scheme. The conservation auction described in this book applies a performance-based, uniform pricing rule. Successful bidders are paid a specified price for each unit of output (or input) they supply (plus any promised additional financial and in-kind payments that were not included in the prespecified price range). That price is the same for all sellers and all levels of output (or input). It is specified by matching information on supply (generated through the conservation auction) with information on demand (generated through economic valuation methods) (see Chapter 4). The result is a price that approximates the efficient price of a pseudomarket established through the PES scheme (see Chapter 6). The bidders are told that they will be offered to supply the quantity of output (or input) they each bid at the efficient price ("offered as bid"). This pricing rule offers the sellers the opportunity to earn a profit. This provides a strong incentive to become a seller of output (or input).

This auction design mimics market principles. Each bidder is a "price taker" who chooses output (or input) quantities given a range of potential "market" prices. This provides each bidder with the opportunity to offer the quantity of output that maximizes individual profit. The incentives to seek informational rent are minimized. Any quantity of output (or input) bid that incurs a cost that is higher (or lower) than the offered price prevents profit maximization. Profit could be increased by decreasing (or increasing) the quantity of output (or input) offered. A design that is based on market principles also ensures that the profit enjoyed by the low-cost sellers is larger than that enjoyed by the high high-cost sellers. This may encourage ongoing investments in innovative, lower-cost production processes to produce output (or input) at minimum cost.

Whether or not bidders become sellers is a self-selecting process that is driven by individual incentives as is observed in competitive markets for goods and services. Each member of the identified pool of prospective sellers (see Section 2.6) has the opportunity to participate in the auction and choose the quantity of output (or input) they each want to supply at the "market" price. If that price is lower than their marginal costs of supply, they will choose to not submit a bid. Only bidders whose costs are smaller than the "market" price will choose to become sellers. Such a self-selecting mechanism is socially inclusive and is likely to be perceived as "fair." It addresses, at least to some degree, concerns about social conflicts that may arise as a consequence of providing alternative income opportunities to only a subset of community members. This is relevant especially in contexts with limited income opportunities and high levels of poverty.

The conservation auction described here applies a single-round, sealed bid approach. The sellers each submit a single bid and do not receive information on the bids of their competitors. In some cases, it might be more efficient to engage seller groups rather than individuals. For example, engaging self-organized groups allows for some flexibility in the allocation of labor, which might reduce the costs of supply. If bidders are groups rather than individuals, it needs to be decided who is authorized to bid on behalf of the group and what safeguards ensure that all group members have given their consent. The bidding mode is another element that has to be chosen with caution as it may have a significant effect on the bidding results. Are sellers requested to submit their bids in person, per mail, over the phone or online? It is crucial that the bidding mode is customized to the specific context.

Which mode is feasible? Which mode generates the least transaction costs for the bidders and the broker? Which mode maximizes competition by minimizing opportunities for collusion?

Competition among the bidders is essential for the effectiveness of a conservation auction. A competitive conservation auction provides the bidders with an incentive to reveal their "true" marginal opportunity costs of supplying an output (or input) net of any personal enjoyment they may experience. However, competition may also be a driver of social conflict. It is crucial that the risk of potential conflicts caused by competition is assessed, mitigated, and managed through an effective community engagement process.

The degree of competition depends on a range of factors that are correlated with each other. Generally, the degree of competition increases with the number of bidders. The critical mass of bidders required for a conservation auction to be competitive depends on the magnitude of opportunity cost differences across bidders. The smaller these differences are, the more bidders are required. Whether or not existing differences are revealed through the conservation auction depends on the extent of opportunities for collusion, which in turn can be influenced by the bidding format and procedures. Formats and procedures that enable bidders to communicate with each other provide opportunities for collusion. For example, online bidding conducted over several weeks involving sellers from a single village would provide ample opportunities for collusion. The level of competition and the extent to which bidders reveal their "true" marginal costs may also be reduced through anchoring effects related to experience with previous payments or knowledge of payments in other regions. Bidders may anchor their bids on such experience and knowledge.

The broader community may supply one or more inputs relevant to the production process of an environmental service. However, the transaction costs of quantifying these inputs and their associated costs may outweigh potential efficiency gains. Their inclusion might reduce the overall PES scheme efficiency.

The Lao PES scheme

An input-based conservation auction was used to estimate the marginal costs of anti-poaching patrolling. The patrol teams were invited to bid for 3-year patrol contracts. All bidders were briefed extensively during training sessions (see Section 5.1) on the terms and conditions of engaging in the antipoaching patrol scheme as specified in the patrol contracts (see Section 7.2 and Annex 7). The terms and conditions specify the patrol commitments, roles, and responsibilities (see Section 2.2), the benefits to the patrol teams (see later), the payment schedule and transfer mechanism (see later), the monitoring and penalty system (see Section 7.2), and the mechanism for grievance, conflict resolution, and redress (see Section 5.1).

The benefits enjoyed by the patrol teams include payments for performing antipoaching patrols, bonus payments for dismantling poacher camps and snare lines, equipment (see Section 5.2), health and accident insurance covering their patrolling activities, and recognitions ("trusted wildlife guardian") for fulfilling their obliga-tions. Patrol teams were told that they may be eligible for rewards from the GoL for activities that support the enforcement of Lao legislation on wildlife protection.

A three-monthly payment frequency with an initial 25% up-front payment at the time of the first patrol was deemed a "good" compromise between the sellers' request for immediate payments and the necessity to minimize transaction costs associated with payment transfers. The payment transfer is organized as follows: the patrol manager calculates the payments every 3 months based on patrol team performance before sending a payment request to the PAFO. The patrol teams are encouraged to calculate their own payments to cross-check the calculations made by the patrol manager. PAFO then transfers the requested amount directly into bank accounts of each patrol team held by a district bank. The patrol teams with-draw money from the team's bank account through a check, which needs to be signed by all team members to avoid fraud.

In addition to the benefits offered to the teams, the PES scheme promised pay-ments to the development funds of the participating villages in return for complying with wildlife conservation agreement. These payments consist of a fixed and a var-iable component made yearly and three-monthly, respectively. The extent of the fixed component was based on a fixed amount per household. It is therefore proportional to the number of households in each village. The extent of variable component is pro-portional to the patrol payments received by the teams of each village (5% of patrol payments). The more patrols teams perform the higher are the variable payments to the respective VDF. Additional benefits eligible to the communities include a village recognition ("trusted wildlife guardian") for fully honoring the commitments set out in the community conservation agreement. The communities were also told that they may be eligible for rewards from the GoL for activities that support the enforcement of Lao legislation on wildlife protection (for example, the provision of evidence that leads to an arrest).

The payment transfer to the VDF is organized as follows: The patrol manager calculates the payments every 3 months based on the performance of the patrol teams of each village (for the variable payments) before sending a payment request to Provincial Office of Natural Resources and Environment (PONRE). PONRE then transfers the requested amount directly into the bank account of the VDF held by a district bank. Withdrawal from the bank account will follow the procedures of the VDF.

The auction applied a single-round, sealed-bid bidding format. Patrol teams each submitted a single bid without knowing the bids of their competitors. Each bid stated the number of patrols each team would be willing to perform per year (for 3 years) at each of the six prespecified price levels (see Fig. 5.8). The endpoints of the price range were based on the PES scheme architects' best guesses of the teams' lowest and highest opportunity costs of patrolling. The lowest price level was represented by the wage for hired labor in a rice paddy. The income opportu-nities attainable in the private sector were used to set the highest price level.

'Busy' season		'Quiet' season	
Price per patrol per team (US$)	Number of patrols per team per year	Price per patrol per team (US$)	Number of patrols per team per year
171	0	171	2
257	1	257	4
342	3	342	6
428	4	428	8
514	5	514	9
599	6	599	10

FIGURE 5.8

Example bidding forms.

Source: From Scheufele G. & Bennett J. (2018) 'Costing biodiversity protection: Payments for Environmental Services schemes in Lao PDR', Vol 7, pp.386-402. ©2018 Journal of Environmental Economics and Policy Ltd.

The data collected during the community consultations indicated that the opportunity costs of patrolling would differ across the year. They were expected to be higher during the "busy" rice planting and harvesting season (4 months) and lower during the "quiet" season (8 months).To capture these expected opportunity cost differences the teams were asked to submit a separate bid for the "busy" and the "quiet" season.

The auction employed a uniform pricing rule. The teams were informed that the price paid per patrol (for all patrols offered) would be one of the six prespecified prices. It was explained that this price would be determined by matching demand with supply. The teams were also informed that they would be offered the number of patrols they each bid at the determined price. Whether or not teams would become sellers of patrols was therefore a self-selective process. Each team decided how many patrols (if any) they wanted to perform at the determined price. The offered opportunity to earn a profit was a strong incentive to engage in patrolling. In addition to the patrol payments, the patrol teams were offered bonus payments for removing snare lines and dismantling poacher camps, a travel allowance to commute between their homes and the PCPPA, a health and accident insurance covering patrolling activities, and social recognition. The teams were told that all equipment they would need for the patrolling activities would be provided by the PES scheme. The teams might have factored this additional benefits into their bids. This might have partially offset their costs and thus increased the number of patrols offered at each price level.

The teams had to submit their bids in person during a bidding session facilitated by an auctioneer. During the bidding, teams were not allowed to communicate with each other. The completed bidding forms had to be submitted in a sealed envelope. Team members were allowed to bid on behalf of the team to account for time constraints of other team members. However, the team bids only became valid once all team members had signed the bidding forms to confirm their consent.

The main challenges of conducting the conservation auctions included the low education levels of the bidders, the risk of collusion, and concerns about fairness. Low education and limited market exposure of the bidders were addressed by an intensive information and training process (see Section 5.1) that proceeded the actual bidding. This process included a detailed review of the patrol scheme and a complete review of the patrol contract. As the teams were taught to factor into their bids all the costs incurred by performing patrols, including those associated with participating in the conservation auction, any transactions costs might therefore be accounted for in the teams' marginal cost of patrolling.

Transaction costs faced by both the teams and the brokers were kept as small as possible by "grouping" the training, information, and bidding sessions.

The opportunities for collusion were minimized by limiting communication opportunities between teams. Even though it was not feasible to conduct the auctions simultaneously in all villages, collusion among villages was made difficult by sequencing the auctions across the eight villages with only a few days in between. Opportunities for collusion were therefore minimal, especially since the bidding took place during the wet season when transport between villages is difficult. However, with the exception of one village, limited telecommunication was possible. Potential collusion was further minimized by stressing the competitive nature of the auctions, placing teams apart from each other, and prohibiting communication across teams.

Potential concerns about fairness were addressed by the applied self-selection mechanism and uniform pricing rule. The former made the participation process socially inclusive. The latter provided the teams with the opportunity to earn a profit, whereby low-cost teams were offered more patrols earning larger profits than high-cost teams. Put simply, teams with the least income and employment opportunities had the opportunity to earn the largest profits.

The auctioneer received bids from all of the eight villages that were invited to participate in the conservation auction. Yet, some of the bids were not valid since they lacked the signature of all team members. These teams were given the opportunity to obtain all the required signatures. To prevent potential bid modifications and associated conflicts, the project team took photographs of the original bid forms. Teams who lost members after the bidding were permitted to replace them with individuals from the same village. In total, 55 teams from all eight target villages submitted valid bids. As shown in Fig. 5.9, the number of bidding teams differed across the villages.

The obtained array of price–quantity pairs revealed the teams' individual marginal cost functions (see Fig. 5.10). The number of patrols offered was a function of price. The higher the price the more patrols were offered. Although the price range was the same for all teams, the number of patrols offered at each price level differed across teams and villages. The observed differences in opportunity costs across villages might be explained by differences in their proximity to roads and the associated access to markets and the availability of alternative income and employment opportunities. The teams from villages located close to the main road only offered patrols at the highest price level, whereas teams from more remote villages also bid at the lower price levels. At any given price, more patrols were offered for the "quiet" season than the "busy" season. This clearly indicated that the opportunity

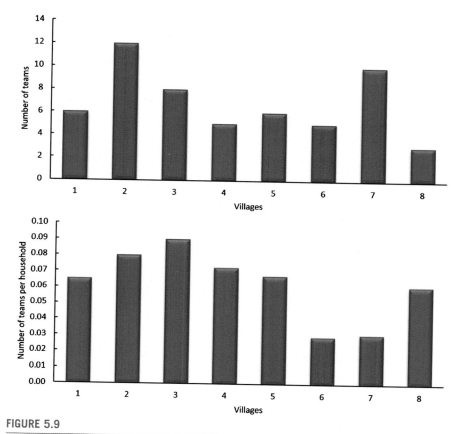

FIGURE 5.9

Team participation by village.

Source: From Scheufele G. & Bennett J. (2018) 'Costing biodiversity protection: Payments for Environmental Services schemes in Lao PDR', Vol 7, pp.386-402. ©2018 Journal of Environmental Economics and Policy Ltd.

costs of patrolling differed not only across teams and villages but also across seasons.

The inputs in the production process supplied by the broader community (see Section 2.6) were ignored. The transaction costs of estimating the associated costs and benefits were deemed to outweigh potential efficiency gains.

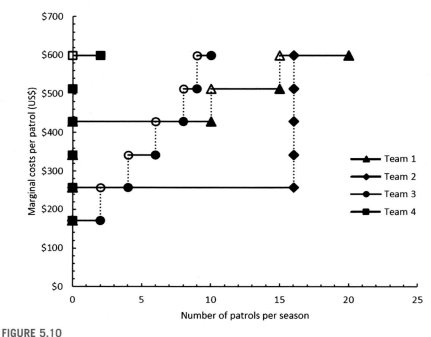

FIGURE 5.10

Sample of marginal costs of patrol team employment at the team level ('busy' season).

Source: From Scheufele G. & Bennett J. (2018) 'Costing biodiversity protection: Payments for Environmental Services schemes in Lao PDR', Vol 7, pp.386-402. ©2018 Journal of Environmental Economics and Policy Ltd.

5.4 Aggregating supply

The estimated individual marginal cost functions need to be aggregated across all sellers to generate information on "market" supply. This is achieved by horizontal summation. Horizontal summation entails adding the quantity of output (or input) all sellers offered to supply at each price. Given that the prespecified price range is discontinuous, both individual and "market" supply are represented by a step function. The number of price levels determines the number of steps and the extent of the "gaps" between price levels.

An input-based conservation auction might not always be able to capture the costs of all inputs used in the production process (see Box 5.1). For instance, the auction may capture the costs associated with planting trees but not those associated with the purchase of seedlings, the purchase of equipment, or the employment of a project manager. Inputs may be "external" to the conservation auction because of limited expertise of sellers to purchase the appropriate equipment, a lack of market access, or a requirement for standardized equipment. In the presence of economies of scale, inputs may be excluded deliberately from a conservation auction to

---

Box 5.1 Costs of production

The **costs of production** are the sum of the opportunity costs of the inputs used in a production process. The costs are classified into **variable costs** and **fixed costs**. Variable costs are the sum of the opportunity costs of all variable inputs used in a production process. Variable costs change with the level of output. Fixed costs are the sum of the opportunity costs of each input that is fixed for a given planning horizon. Fixed costs are independent of the level of output. For example, increasing the in-stream water quality of a river (output) increases the payments made to planters (variable costs), whereas the monthly wage of the coordinator remains constant (fixed costs). The former depend on the number of trees planted and sediment barriers installed, whereas the latter do not.

Sunk costs are the costs of an investment that cannot be recovered by reselling the purchased asset. The opportunity costs of such an investment are zero. Sunk costs are thus irrelevant to decision-making once the investment has been made. For example, the investment in the employment of a water quality manager with a 1-year contract is a sunk cost.

---

reduce costs. For instance, buying seedlings "in bulk" may be cheaper than each seller buying their own. Or, some inputs may involve tasks not assigned directly to the sellers but performed by the broker such as project management tasks. Information on input costs not captured by a conservation auction might be obtained from established markets for goods and services such as the price of equipment or the average salary of a project manager.

Any costs generated through the use of "external" inputs used to produce the output need to be added to the "market" supply function. Are these "external" input costs fixed or variable? Any "internal" input cost estimated through the described conservation auction is variable by definition. Bidders are offered a constant price per additional unit of input, which produces an additional unit of output. An "external" input cost, however, may be variable or fixed. For example, a bonus payment that is linked to an "internal" input but not captured in a conservation auction would be considered an "external" variable cost. In contrast, the "external" cost associated with the employment of a project manager would be classified as fixed. It remains the same, independent of the quantity of output produced. Actually, it would be considered a sunk cost and would need to be added fully to the first step of the "market" supply function.

However, classifying "external" costs as fixed or variable may not always be straightforward. In fact, an "external" input cost may be both variable and fixed depending on the definition of the "marginal unit" used as a reference. At each step of the "market" supply function, additional sellers enter the market. The recruitment of additional sellers may involve "external" input costs such as the cost of purchasing equipment. These costs would be fixed with respect to the quantity of "internal" input within each step but variable with respect to the "external" input across steps. The "external" input cost would be independent of the output quantity within each step (the margin refers to the "internal" input unit) but dependent across steps (the margin refers to the "external" input unit). For example, the cost of shuffles depends on the number of sellers who are willing to plant trees at a certain price level but not on the number of trees they each plant at that level. The applied "offered-as-bid" pricing rule justifies averaging the "external" input cost across all "internal" input units within each step.

For some temporal settings, adding an "external" cost to the "market" supply function would require its distribution across the length of the supply period. The time period used to estimate "market" supply may differ from the time period of seller engagement formalized in a legally binding contract. For example, sellers may be asked to bid the quantity of an input they are willing to supply per year over the course of 3 years. "Market" supply would be estimated per year. Yet, the equipment would be used for 3 years.

As discussed in this section, adding "external" input costs to the "market" supply function might be a challenging and costly task. The decision of excluding inputs from a conservation auction should therefore be made with caution. It should be an informed decision accounting for the extent of all transaction costs involved.

Lao PES scheme

The team supply functions estimated through conservation auctions were aggregated to generate a "market" supply function for each season (see Table 5.1 and Fig. 5.11).[3] This was achieved by adding the number of patrols offered by all teams at each price level.

Any input costs that were not accounted for in the bids ("external costs") were added to the "market" supply functions. These included costs associated with the purchase of insurance and patrol team equipment, bonus payments for snares

Table 5.1 Overall marginal costs at the 'market' level.

Price level (US$)	'Busy' season		'Quiet season'	
	Overall marginal costs (US$)	Number of patrols	Overall marginal costs (US$)	Number of patrols
$171	$639	14	$532	39
$257	$334	79	$323	255
$342	$382	177	$388	432
$428	$466	261	$471	574
$514	$544	348	$544	664
$599	$648	428	$643	775

Source: From Scheufele G. & Bennett J. (2018) 'Costing biodiversity protection: Payments for Environmental Services schemes in Lao PDR', Vol 7, pp.386-402. ©2018 Journal of Environmental Economics and Policy Ltd.

---

[3] The bids, and therefore, the marginal costs might have been offset, to some extent, by the benefits associated with the insurance cover and the recognitions provided to the patrol teams.

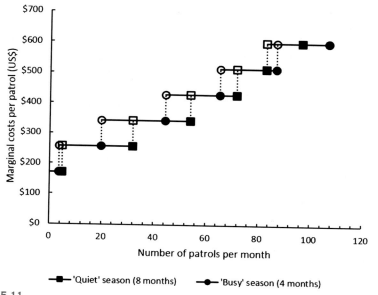

FIGURE 5.11

'Market' marginal costs of patrol team employment by season.

Source: From Scheufele G. & Bennett J. (2018) 'Costing biodiversity protection: Payments for Environmental Services schemes in Lao PDR', Vol 7, pp.386-402. ©2018 Journal of Environmental Economics and Policy Ltd.

removal and camp dismantling, the employment of a patrol manager, and variable payments to the VDFs (see Fig. 5.12). The fixed payments to the VDFs were not added to the "market" supply functions. This is justified since the benefits generated through these payments were not added to the "market" demand function either. Quantifying these benefits was deemed too costly. Limited expertise on the part of the teams, the requirement for standardized uniforms, and economies of scale were the main reasons for providing the equipment through the PES scheme. A lack of market access and economies of scale were the drivers of externalizing the purchase of the insurance.

Information on input costs associated with the purchase of equipment, insurance and patrol manager employment were obtained from established markets for goods and services. The equipment was purchased to be used over the 3 year contract period. Annual equipment costs were calculated as average costs over 3 years. Repair and replacement costs were not included in this calculation. Expert opinion[4] was the basis for determining the extent of the bonus payments for dismantling camps and the variable payments to the VDFs. A combination of expert opinion (price per snare

---

[4] The expert survey used to inform the development of the model is presented in Annex 1.

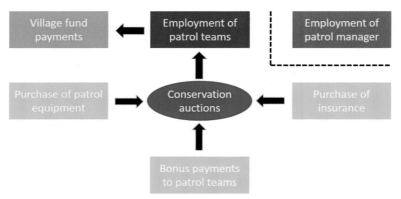

FIGURE 5.12

Cost structure.

Source: From Scheufele G. & Bennett J. (2018) 'Costing biodiversity protection: Payments for Environmental Services schemes in Lao PDR', Vol 7, pp.386-402. ©2018 Journal of Environmental Economics and Policy Ltd.

wire) and a stochastic model (predicted number of snare wires collected per patrol) was used to estimate the cost of the bonus payments for snare lines removal. Since the bonus payments depend on snare line densities they decrease with an increasing number of patrols. They might need to be adjusted to maintain the desired incentives.

Marginal units of the "external costs" differed across inputs. The costs of the bonus payments and the payments to the VDFs were calculated on a per additional patrol" basis. The costs associated with the insurance and the equipment provided to the teams were calculated as a cost per additional team", while the costs of equipment shared among teams within each village were calculated on a per additional village" basis. At each step of the "market" supply function, additional teams enter the market. The costs per additional team" and per additional village" were fixed with respect to the number of patrols within each step but variable with respect to the number of teams and villages across steps. Adding any "external costs" calculated on a per additional team" and per additional village" basis to the "market" supply function required their conversion into a cost per additional patrol". This entailed dividing the costs of all additional teams and villages by the total number of additional patrols employed at each price level. The dual character of the conservation auction required distributing these costs across the "busy" and the "quiet" season. "Equivalent" quantities had to be calculated to account for differences in bidding behavior across teams. At each price level, additional teams either offered patrols for both seasons or the "quiet" season only. The cost of employing a patrol manager was classified as fixed and sunk and added to the first price level. This cost was calculated per additional PES scheme" and converted into a cost per additional patrol" as an average cost.

"Market" supply including all "external costs" and "market" supply disaggregated in "internal cost" and "external costs" are shown in Figs. 5.13 and 5.14, respectively.

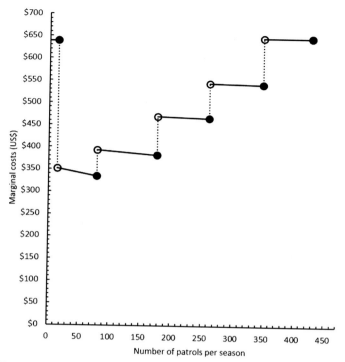

FIGURE 5.13

Overall marginal costs at the 'market' level ('busy' season).

Source: From Scheufele G. & Bennett J. (2018) 'Costing biodiversity protection: Payments for Environmental Services schemes in Lao PDR', Vol 7, pp.386-402. ©2018 Journal of Environmental Economics and Policy Ltd.

5.5 **Challenges and limitations**

A lack of competition is expected to limit the effectiveness of a conservation auction. If the number of bidders falls below a critical threshold a conservation auction fails to capture opportunity cost differences among sellers. In some cases, producing the desired environmental service might require to involve all sellers. For example, all farmers of a ground water catchment might have to reduce their fertilizer inputs to achieve strict water quality standards. This would require to set the price range such that all farmers have an incentive to participate and reduce their respective fertilizer inputs to the extent necessary to meet the standard.

Potential market barriers are further challenges that need to be addressed. Otherwise, these barriers might restrict seller recruitment. Sellers might be excluded from a conservation auction because of limited experience with markets or low levels of literacy. However, they might be the sellers with the lowest opportunity costs of supply. Excluding them would decrease competition among sellers, which would reduce

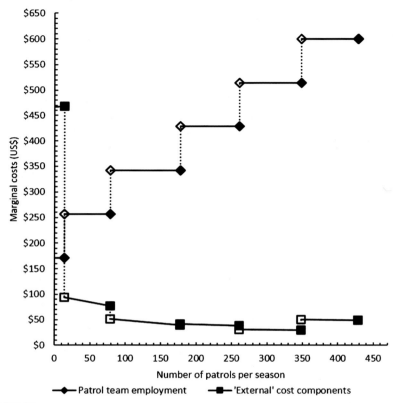

FIGURE 5.14

Marginal costs at the 'market' level disaggregated by 'internal cost' and 'external costs'.

Source: From Scheufele G. & Bennett J. (2018) 'Costing biodiversity protection: Payments for Environmental Services schemes in Lao PDR', Vol 7, pp.386-402. ©2018 Journal of Environmental Economics and Policy Ltd.

PES scheme efficiency. Additionally to efficiency concerns, involving low-income sellers might be desirable if poverty reduction is a secondary PES scheme goal. In any case, customized bidding training may be an effective way to remove some of these barriers and address concerns of social exclusion.

PES schemes that encourage the supply of multiple environmental services pose particular challenges. Multiple conservation auctions may be required to generate separate "market" supply functions for each environmental service. The transaction costs associated with this task are expected to increase with an increasing number of environmental services. The transaction costs may be increased further if the broker has to estimate the marginal costs of supplying input rather than output. This is especially the case if some of the inputs are "external" to the conservation auction. If producing one or multiple environmental services involves more than one "internal" input multiple conservation auctions have to be conducted to estimate a separate "market" supply function for each input. These functions can be added horizontally only if the same price range was used across all inputs. However, setting a common price range effective in capturing opportunity cost differences across both sellers and inputs might not be feasible if the opportunity cost differences across inputs are larger than those across sellers. The common price range might reveal differences across inputs, whereas any differences across sellers might be comparatively too small to be revealed.

References and further readings

Special acknowledgement is given to Xiong Tsechalicha, Saysamone Phoyduangsy and Yiakhang Pangxang for collecting data and conducting the community consultations and conservation auctions.

Special acknowledgement is also given to Phouphet Kyophilavong for coordinating these efforts.

Publications of team members and collaborators of the Lao PES project:

Scheufele, G., Bennett, J., 2018. Costing biodiversity protection: payments for environmental services schemes in Lao PDR. Journal of Environmental Economics and Policy 7, 386—402.

Scheufele, G., Bennett, J., 2017. Can payments for ecosystem services schemes mimic markets? Ecosystem Services 23, 30—37.

Scheufele, G., Bennett, J., 2017. Research Report 16: Costing Biodiversity Protection: Payments for Environmental Services Schemes in Lao PDR. Crawford School of Public Policy, Australian National University, Canberra.

Scheufele, G., Bennett, J., 2013. Research Report 1: Payments for Environmental Services: Concepts and Applications. Crawford School of Public Policy, The Australian National University, Canberra.

Scheufele, G., Bennett, J., Kragt, M., Renten, M., 2014. Research Report 3: Development of a 'virtual' PES Scheme for the Nam Ngum River Basin. Crawford School of Public Policy, The Australian National University, Canberra.

Tsechalicha, X., 2017. Research Report 14: Engaging Communities in a Payments for Environmental Services Scheme for the Phou Chomvoy Provincial Protected Area. Crawford School of Public Policy, Australian National University, Canberra.

Publications of authors external to the project:

Barbier, E., Hanley, N., 2009. Pricing Nature: Cost-Benefit Analysis and Environmental Policy. Edward Elgar PublishingPublishing, Cheltenham.

Coggan, A., Whitten, S.M., Bennett, J., 2010. Influences of transaction costs in environmental policy. Ecological Economics 69, 1777—1784.

Eigenraam, M., Strappazzon, L., Lansdell, N., Ha, A., Beverly, C., Todd, J., 2005. Ecotender: Auction for Multiple Environmental Outcomes. Department of Primary Industry, Melbourne.

Engelbrecht-Wiggans, R., 1980. Auctions and bidding models. Management Science 26, 119—121.

Ferraro, P.J., 2008. Asymmetric information and contract design for payments for environmental services. Ecological Economics 65, 810—821.

Frank, R., 2003. Microeconomics and Behavior. McGraw-Hill, Boston.

Holt, C., 1980. Competitive bidding for contracts under alternative auction procedures. Journal of Political Economy 88, 433—445.

Kagel, J.H., Roth, A.E., 1995. The Handbook of Experimental Economics. Princeton University Press, New York.

Klemperer, P., 1999. Auction theory: a guide to the literature. Journal of Economic Surveys 13, 227—286.

Latacz-Lohmann, U., van der Hamsvoort, C., 1997. Auctioning conservation contracts: a theoretical analysis and application. American Journal of Agricultural Economics 79, 407—418.

Latacz-Lohmann, U., Van der Hamsvoort, C., 1998. Auctions as a means of creating a market for public goods from agriculture. Journal of Agricultural Economics 49, 334—345.

Lopomo, G., Marx, L.M., McAdams, D., Murray, B., 2011. Carbon allowance auction design: an assessment of options for the United States. Review of Environmental Economics and Policy 5, 25—43.

Martínez-Alier, J., 2004. The Environmentalism of the Poor: A Study of Ecological Conflicts and Valuation. Oxford University Press, New Delhi.

Mankiw, N.G., 1998. Principles of Economics. The Dryden Press, Harcourt Brace College Publishers, Fort Worth.

Mas-Colell, A., Whinston, M., Green, J., 1995. Microeconomic Theory. Oxford University Press, Oxford.

McAfee, R.P., McMillan, J., 1987. Auctions and bidding. Journal of Economic Literature 25, 699—738.

Muradian, R., Corbera, E., Pascual, U., Kosoy, N., May, P.H., 2010. Reconciling theory and practice: an alternative conceptual framework for understanding payments for environmental services. Ecological Economics 69, 1202—1208.

Narloch, U., Drucker, A.G., Pascual, U., 2017. What role for cooperation in conservation tenders? paying farmer groups in the high Andes. Land Use Policy 6, 659—671.

Narloch, U., Pascual, U., Drucker, A.G., 2013. How to achieve fairness in payments for ecosystem services? Insights from agrobiodiversity conservation auctions. Land Use Policy 35, 107—118.

Pascual, U., Muradian, R., Rodríguez, L.C., Duraiappah, A., 2010. Exploring the links between equity and efficiency in payments for environmental services: a conceptual approach. Ecological Economics 69, 1237—1244.

Rolfe, J., 2016. Using auctions for conservation: the Australian experience. In: Bennett, J. (Ed.), Protecting the Environment, Privately. World Scientific Press, London, pp. 253—272.

Tietenberg, T., Lewis, L., 2009. Environmental and Natural Resource Economics. Pearson International Edition, Boston.

UN-REDD (United Nations Collaborative Programme on Reducing Emissions from Deforestation and Forest Degradation in Developing Countries), 2013. Guidelines on Free, Prior and Informed Consent. FAO, UNDP, UNEP.

Vatn, A., 2010. An institutional analysis of payments for environmental services. Ecological Economics 69, 1245—1252.

Vickrey, W., 1961. Counterspeculation, auctions and competitive sealed tenders. The Journal of Finance 16, 8—37.

Vickrey, W., 1962. Auctions and Bidding Games: Recent Advances in Game Theory. Princeton University Press, Princeton.

Making the "market"—bringing buyers and sellers together

6

In Chapter 4, the buyer side of the Payments for Environmental Services (PES) scheme equation was detailed. Methods used to estimate the "demand" held for environmental services were detailed and the process used to aggregate "willingness to pay" across segments of the community and over the varying time frames involved was set out. This enables the PES scheme analyst to put together a picture of how much people want the environmental services on offer from a PES scheme. In particular, the relationship between the amount prospective buyers are willing to pay for more environmental services and the amount of environmental services already available can be delineated using the information generated. This is the so-called demand curve for environmental services.

In Chapter 5, the seller side was considered. Ways to determine what potential sellers are "willing to accept" for their provision of environmental services or actions that would produce environmental services were outlined. The combination of various elements of the costs associated with the production of environmental services, especially where different parties are involved in specialist production tasks, was demonstrated to be potentially complex. Aggregation across the cost elements to produce a relationship between willingness to accept and the amount of environmental services provided or actions performed produces what is known as the supply curve.

PES schemes require the bringing together of buyers and sellers in a pseudomarket context. This means that the information on the demand curve and that relating to the supply curve must be integrated. Crucially, this integration allows the determination of the price to be paid by buyers for environmental services and the amount sellers receive for supplying these services or the actions they take. The price must work to coordinate the buyer side of the PES scheme with the seller side simply because the price paid must equal the price received for all parties to be satisfied (see Box 6.6).

The remainder of the chapter sets out the determination of the price paid to sellers for supplying environmental services or actions and the price to be paid by buyers for the environmental services they consume. Matching payments from buyers to those received by sellers is part of this process. Acknowledging and incorporating the transaction costs involved in the PES scheme itself is integral to that process.

Buying and Selling the Environment. https://doi.org/10.1016/B978-0-12-816696-3.00006-5

Box 6.1 Price setting

Price acts to coordinate the actions of buyers and sellers in a market. If the price is too high, there will be more people wanting to sell than to buy. If it is too low, there will be an excess of buyer interest. Price changes in a competitive market to ensure that the plans of sellers to produce exactly match the plans of buyers to consume. When the price is too high, sellers will be encouraged to produce a lot while buyers are less keen to make purchases. The excess supply that emerges forces prices down as sellers compete with each other to get rid of their excess stocks. In contrast, a price that is too low will result in shortages as buyers scramble to secure bargains while sellers are not so interested to produce. The competition between buyers pushes the price up.

The net benefit enjoyed by the buyers is the difference between what people are willing to pay (their marginal benefit) and what they have to pay (the price). This so-called "consumer surplus" is a measure of what buyers see as their benefits net of their costs. For the sellers, the equivalent measure of net benefit is the difference between their costs of producing more (their marginal costs) and the price they receive for the product. This is called the "producer surplus." Together, the consumer and producer surpluses make up the total addition to society's well-being that results from the exchanges between buyers and sellers.

The goal of a PES scheme seeking to maximize the well-being of people is to set a price for environmental services that equates supply with demand. That price is said to be efficient because any other price would deliver a lower level of well-being for buyers and sellers.

6.1 Determining an efficient "market" price

The efficient "market" price per unit of an environmental service paid by buyers and received by sellers is determined by equating the aggregate marginal cost of "market" supply (see Section 5.4) with the aggregate marginal benefits driving "market" demand (see Section 4.5). This "negotiated" price is uniform (all buyers and sellers face the same price) and constant across supply (the price is independent of the output amount) (see Section 4.5). Equating supply with demand also determines the efficient amount of environmental service supply. If sellers bid to perform actions (inputs) instead of supplying an environmental service (output), the demand for output needs to be converted into a demand for inputs (see Sections 3.2 and 4.5). Alternatively, input supply has to be converted into output supply. The efficient amount is the sum of the output (or input) amounts each prospective seller bid to supply at the efficient price in the conservation auctions (see Section 5.3).

Once the amount of total output (or input) has been determined, each prospective seller is offered a contract (see Section 7.2) to supply the output (or input) amounts they each bid at the efficient price (self-selection). By design, this allocation mechanism is based on self-selection, and therefore socially inclusive (see Section 5.3). It ensures that low-cost sellers are offered to supply a larger amount of output (or input) than high-cost sellers. Given the same price is paid for all units supplied this mechanism also guarantees that low-cost sellers earn larger profits than high-cost sellers. The efficient quantity of output (or input) is supplied at the lowest possible costs. The waste of scarce resources is minimized: the scheme is cost-effective.

Low-production costs typically signal low-opportunity costs due to limited income and employment opportunities. That means the poorer segments of society benefit disproportionally from PES scheme participation. Furthermore, the allocation mechanism used ensures that both buyers and sellers are made better off. It thus addresses equity concerns, especially in developing countries, where the buyers of environmental services are typically wealthier than those who are willing and able to supply them.

Not all sellers may accept the contracts they are offered. Contract rejections may be explained by a change of sellers' opportunity costs after the bidding. For example, a prospective supplier may be offered another source of employment in the period between when they made their bid (when their opportunity costs were relatively low) and when an offer was made (when their opportunity costs had increased). Minimizing the time span between the conservation auctions and the announcement of contract offers is likely to reduce the number of contract rejections. More contract rejections could be expected in the context of a rapidly developing economy where the employment/business opportunities available to people are expanding dramatically.

Supplying less than the efficient amount of output (or inputs) because of contract rejections diminishes a scheme's efficiency. This can be addressed by offering the surplus of demand to the sellers who accepted their offers at the prices they bid. To maintain social cohesion, it is crucial that any such reallocation process is transparent and perceived to be fair by all stakeholders involved.

The Lao PES scheme

The aggregate marginal benefit (demand) in "input space" associated with the environmental service KNOWLEDGE (see Section 4.5) was equated with the aggregate marginal cost (supply) associated with antipoaching patrolling (see Section 5.4) to estimate the efficient price per patrol and team for each season. The relevant marginal benefits and costs are listed in Table 6.1. It was assumed for the price setting exercise that the contracted sellers would fully comply with their contracts. As complete contract compliance is unlikely, some loss of efficiency is to be expected as a result of making this assumption. Supply equals demand at the efficient price (p^e) and efficient quantity (q^e) (see Fig. 6.1). The efficient price was found to be the same for the "busy" and the "quiet" season. This suggests that the differences between the prespecified price levels used in the auctions are larger than any marginal cost difference across the two seasons. The difference between seasonal marginal costs is picked up solely by a difference in the season-specific efficient number of patrols. This implies that the amount of patrol teams willing to work in the "busy" season is smaller than that for the "quiet" season. This seasonal imbalance in patrol effort may provide incentives for poachers to focus their activities more on the "busy" season. Consequently, the bioeconomic modeling predictions may exaggerate the benefits generated in the "busy" season. This outcome could be rectified by reducing the price "gaps" between the price levels to be used in subsequent

Table 6.1 Market demand and supply in 'input' space.

Number of patrols per season	Marginal benefits	Marginal costs
Busy season		
42	$399	$298
124	$19	$108
252	$3	$113
354	$1	$133
435	$1	$157
506	$0	$182
Quiet season		
117	$399	$180
385	$19	$100
671	$3	$111
893	$1	$133
1099	$1	$157
1337	$0	$182

Source: Scheufele G., Bennett J. & Kyophilavong P. (2018) 'Pricing biodiversity protection: Payments for Environmental Services schemes in Lao PDR', Land Use Policy, Vol 75, pp. 284–291.

conservation auctions. Increasing bidding price points would however increase the cognitive burden faced by the bidding teams so that changing the auction arrangement would need to be thought through carefully.

The number of patrols bid by the prospective sellers at the efficient price was found to exceed the efficient number of patrols for both seasons. This is a direct result of the discontinuous price points used in the conservation auctions (see Section 5.3). The number of price levels determines the number of steps and the extent of the "gaps" between price levels. The "step function" character of the supply function, together with the promise made that each team would be offered the number of patrols they each bid at the announced efficient price (see Section 5.3), create some complexities. The PES scheme designer had to decide whether to contract the number of offered patrols at the efficient price (fourth price level) or that offered at the next lower price level (third price level). The former would have meant that the number of patrols contracted would be beyond the efficient amount. Offering at the lower price level would have meant contracting fewer patrols than the efficient level (see Fig. 6.1). The two scenarios and their respective outcomes are discussed in the next section.

6.2 Determining social net benefit

Determining the social net benefit generated through the supply of environmental services enables an assessment of a PES scheme's efficiency of and effectiveness in generating additional environmental services. Social net benefit signals the extent

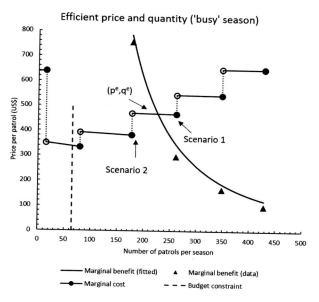

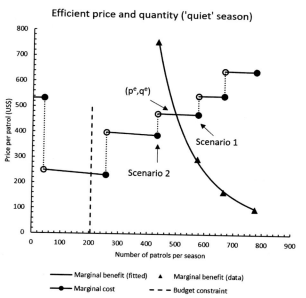

FIGURE 6.1

Efficient price and quantity in 'input' space.

Source: Scheufele G., Bennett J. & Kyophilavong P. (2018) 'Pricing biodiversity protection: Payments for Environmental Services schemes in Lao PDR', Land Use Policy, Vol 75, pp. 284–291.

by which buyers and sellers are made better off through their engagement in the scheme. The respective benefits they each enjoy are measured in terms of consumer and producer surplus, respectively. Consumer surplus is estimated by subtracting the aggregate buyer payment from the aggregate of the marginal benefits of the environmental services enjoyed by the buyers. Producer surplus is approximated by subtracting the aggregate of the marginal costs of supply from the payments that sellers receive.

Buyers and sellers face the same price per unit of environmental service (output) or actions (inputs). Consequently, the aggregate payments made by the buyers equal the aggregate of payments made to sellers. This amount is calculated by multiplying the efficient amount of output (or input) with the efficient "market" price per unit of output (or input). An environmental service may be purchased by more than one buyer group. In this case, the groups' respective willingness to pay amounts can be used as weights to split, proportionally, the aggregate payment amount between the buyer groups.

Acknowledging and assessing transaction costs is an integral part of evaluating whether a PES scheme makes society better off (i.e., if the scheme creates a net social benefit). Some of the transaction costs borne by buyers, sellers, and agents may be captured by the supply and demand estimations. For example, prospective sellers may incorporate their cost of participating in the conservation auctions into their bids. Similarly, buyers may have accounted for their transaction costs when expressing their willingness to pay. Similarly, some of the transaction costs borne by the broker(s) can be incorporated in the estimation of the marginal cost of supply. Such costs may include the "external" costs incurred by the broker in purchasing equipment or the costs associated with the employment of a PES scheme manager. Yet, "up-front" transaction costs associated with the development and initialization of a PES scheme borne by the broker(s) (for example, costs generated in the development of bioeconomic models, conservation auctions, and nonmarket valuation surveys) may not have been accounted for in the social net benefit calculated for the length of the supply contracts. Typically, these "up-front" costs are relatively high. Yet, the extent of these "'up-front" costs relative to the generated social net benefit depends on a scheme's expected overall lifespan (beyond the contract length). Given an expected perpetual social net benefit stream, the costs' relative significance keeps going down as long as the scheme is operational. Note also that once these "up-front" costs are incurred, they become "sunk." In other words, those costs become irrelevant to the calculation of the transaction costs of future iterations of the PES scheme or new PES schemes that use the previously generated information. In principal, however, society is only made better off if the benefits outweigh the costs (including all transaction costs).

The Lao PES scheme

The annual aggregate payment made by the buyers (tourists and residents) and received by the sellers (patrol teams) was calculated by multiplying the efficient number of antipoaching patrols with the efficient price per patrol. The tourists' and residents' respective willingness to pay per patrol were used as weights to split, proportionally, the aggregate buyer payment across the two buyer groups. The amount individual buyers have to pay was calculated by dividing each groups' annual aggregate buyer payment by the number of their respective group buyers. Consumer surplus was calculated by subtracting the annual aggregate buyer payment from the annual aggregate of marginal benefits (willingness to pay) of patrolling enjoyed by the buyers. Producer surplus was calculated by subtracting the annual aggregate of the marginal costs of patrolling from the annual aggregate payments sellers receive.

Some of the supply costs are borne by the brokers (see Section 5.4). These annual "external" input costs (transaction cost) are included in the annual aggregate costs of patrolling. This implies that the efficient price paid by the buyers is higher than price paid to the patrol teams. Consequently, producer surplus had to be disaggregated into "broker" surplus and the surplus secured by the patrol teams. "Broker" surplus was calculated by subtracting the annual "external" input costs from the annual "external" revenue. The "external" revenue was calculated by multiplying the price "paid" to the broker (the difference between the efficient price per patrol paid by the tourists and residents and the price paid to the patrol teams) with the efficient annual amount of patrols. The surplus "enjoyed" by the broker is negative. This is a direct consequence of the auction design that excluded some of the inputs costs ("external" input costs) from the bids. A part of the overall revenue is allocated twice: to cover the patrol payments and the "external" input costs.

The surplus earned by the patrol teams was calculated by subtracting their annual aggregate of their marginal cost of patrolling from the annual patrol team revenue secured through patrol and bonus payments (see Section 5.4). The patrol payments were calculated by multiplying the "efficient" annual number of patrols by the price paid to the patrol teams. The bonus payments were calculated by multiplying a fixed price per snare and camp by the predicted annual amount of snares and camps the teams would collect and dismantle. The predictions were generated by the stochastic population model used to estimate the marginal products of the outputs (see Section 3.1).[1] The patrol teams were assumed to have incorporated the transaction costs of participating in the auctions into their bids, and therefore, into their supply costs. The surplus they enjoy is therefore assumed to be exclusive of these transaction costs.

The outcomes of the two scenarios discussed in the previous section (paying above the efficient price and paying below the efficient price) are presented in Table 6.2. The net benefit generated by the patrol effort associated with the (lower) third price level (scenario 2) is larger than that associated with the (higher) fourth price level (scenario 1). Under scenario 1, the additional costs outweigh the

---

[1] The PES scheme was designed to enable an adjustment of the fixed price per snare and camp to compensate for a decrease in bonus payments as snare and camp density decline over time. Such an adjustment may become necessary to keep the incentives attractive to the patrol teams.

Table 6.2 PES scheme outcomes predicted per year (under scenario 1 and scenario 2).

	Scenario 1		Scenario 2	
Number of villages with patrol teams	7		6	
Number of patrol teams	49		42	
Number of patrols	835		609	
Average patrol frequency per km2 and month	4.6		3.3	
Total benefits	$5,331,865		$5,315,034	
Total costs	$337,594		$231,290	
Total cost to teams	*$272,119*		$175,392	
Total cost to agents	*$65,475*		$55,898	
Total net benefits (total surplus)	$4,994,271		$5,083,744	
Aggregate buyer payment	$391,855	$425,112	$235,344	$267,982
Aggregate payment from urban residents	*$36,186*	$39,257	$21,720	$24,732
Aggregate payment from international tourists	*$355,669*	$385,855	$213,624	$243,250
Aggregate supplier revenue	$391,855		$235,344	
Aggregate supplier revenue to teams	*$359,636*		$212,084	
Aggregate supplier revenue to agents	*$32,219*	$65,475	$23,260	$55,898
Consumer surplus	$4,940,010	$4,906,754	$5,079,690	$5,047,052
Producer surplus	$54,261	$87,518	$4,055	$36,692
Producer surplus to teams	*$87,518*		$36,692	
Producer surplus to agents	*−$33,257*	$0	−$32,637	$0
Variable payments to village development funds	$17,869		$10,426	

Total cost to agents includes variable payments to village development funds.
Source: Scheufele G., Bennett J. & Kyophilavong P. (2018) 'Pricing biodiversity protection: Payments for Environmental Services schemes in Lao PDR', Land Use Policy, Vol 75, pp. 284–291.

additional benefits. For both scenarios, the resident payment is about 10 times smaller than the tourist payment. The main two drivers of this difference are the difference in population sizes and the difference in their respective willingness-to-pay estimates (*see* Section 4.4). It also must be noted that, when there is a relatively large number of patrols, the magnitude of the marginal benefits is largely driven by the number of buyers rather than the marginal products. The marginal products for both environmental service outputs (species and populations) were predicted to be close to (but never equal to) zero (see Section 3.2). This is a direct result of the lack of data that prevented the inclusion of negative impacts of disturbance on wildlife diversity caused by "too many" patrols. It is expected that including these impacts into the bioeconomic model would have resulted eventually in zero or negative marginal products being generated by more patrols, and therefore in zero or negative marginal benefits being generated. This, in turn, would have lowered the efficient price and the efficient number of patrols.

To this point, the analysis of social net benefit has assumed that buyer payments were already available. Yet, until a sustainable funding mechanism is in place (for instance, a levy on tourist visa costs or a surcharge on domestic electricity prices, as suggested by the choice modeling payment vehicles), seed funding provided by World Bank under the Protected Area and Wildlife Project is being used (see Section 2.5) to fund the scheme.

The seed funding available was not sufficient to purchase the efficient amount of patrols (scenario 2). It allowed the purchase of either all of the patrols offered at the (lowest) first price level or 80% offered at the second price level. Purchasing at the first price level would have led to an employment of only about 15% of the bidding teams. Such an outcome was considered to run the risk of being perceived as unfair among the bidding patrol teams, potentially creating social disruption and perverse incentives. Instead, the number of patrols offered at the second price level was reduced by 20% across all 32 teams that bid at that price level.

The budget constraint introduced an additional challenge. It lowered the price paid per patrol to an amount that would have excluded all teams from a specific village that is of strategic importance to the PES scheme's effectiveness (and thus efficiency) as it is the only village located at the northern border of the protected area. A supplementary payment had to be offered to the teams from that village to provide a sufficient incentive for them to participate in the scheme.[2]

Eventually, 17 teams from five villages signed patrol contracts (see Section 7.2). This implied the purchase of about 72% of the patrols that were offered to the teams at the "budget constraint" price. Some of the teams did not accept the offer. One explanation for these rejections may have been rumors of the opening of a new mining operation near the villages and associated higher wages than those offered in the PES scheme. In other words, the teams' opportunity cost of time might have gone up. The loss of patrols was only partially offset by new teams (formed from teams with members who rejected the offer) and teams who bid at a price lower than the one offered. These teams accepted the bids submitted by the teams that withdrew. The outcomes predicted by the purchased number of patrols are summarized in Table 6.3.

---

[2] The supplement payment boosted the price paid per patrol by $85.60 and the variable village payment by $4.28 per patrol.

Table 6.3 PES scheme outcomes predicted per year (purchased patrols).

	Purchased patrols	
Number of villages with patrol teams	5	
Number of patrol teams	17	
Number of patrols	196	
Average patrol frequency per km2 and month	1.1	
Total benefits	5,037,217	
Total costs	$80,999	
Total cost to teams	*$49,647*	
Total cost to agents	*$31,352*	
Total net benefits (total surplus)	$4,956,218	
Aggregate buyer payment	$67,748	$88,018
Aggregate payment from urban residents	*$6188*	*$8040*
Aggregate payment from international tourists	*$61,559*	*$79,978*
Aggregate supplier revenue	$67,748	
Aggregate supplier revenue to teams	*$56,666*	
Aggregate supplier revenue to agents	*$11,082*	*$31,352*
Consumer surplus	$4,969,470	$4,949,199
Producer surplus	−$13,252	$7019
Producer surplus to teams	*$7019*	
Producer surplus to agents	*−$20,271*	*$0*
Variable payments to village development funds	$2568	

Total cost to agents includes variable payments to village development funds.
Source: Scheufele G., Bennett J. & Kyophilavong P. (2018) 'Pricing biodiversity protection: Payments for Environmental Services schemes in Lao PDR', Land Use Policy, Vol 75, pp. 284–291.

6.3 Challenges and limitations

Selling and buying multiple environmental services pose particular challenges to the PES scheme architect. The transaction costs associated with "making" a market multiply if separate prices for separate environmental services (or actions) and buyer groups have to be determined. The more environmental services to be sold and bought, the higher will be the transaction costs.

Yet, the main challenges are expected to be a direct result of practical obstacles. Addressing these challenges requires compromise and flexibility. There is no such thing as a "perfect" PES scheme. The transaction cost involved in generating all the information needed will always outweigh the gains that are achievable by a fully efficient scheme.

Flexibility may be required in redistributing supply allocations if sellers reject offers. The rejection rate may be minimized by offering the contracts as soon as possible after the conservation auctions. Strategic considerations may challenge the allocation mechanism used to select the sellers. Deviating from a strictly uniform

distribution of those payments, the payment mechanism (including where payments are to be made and to whom), how and by whom monitoring of payments will be made, and the penalties associated with failure to comply.

However, buyers are usually the most problematic party to a PES scheme. Because of the open access characteristics of many environmental services, buyers can be difficult to identify, hard to establish their values and even harder to contract to pay. Free riding is a real obstacle to PES scheme implementation. Substantial effort may be required to secure PES buyers, and contracting can be an integral part of the process needed both to identify and then to formalize the receipt of payments.

As identified in Chapter 4, it is possible to search out potential buyers of environmental services and estimate their marginal benefits. In the process of estimating values, payment vehicles can be set up and those can be used as the starting point for contracts with buyers. Even where free riding is pervasive, contracts enforcing payment for environmental services by buyers can be achieved. This involves the addition of an environmental service payment alongside a compulsory payment. In this way, PES payments can be made by "piggy-backing" on another form of payment.

Payment contracts may also be negotiated with organizations that represent buyers. In the broadest context, elected governments can be regarded as buyers' agents that can enter into contracts by being obliged to support a PES scheme from general tax revenue. Even international agencies and NGO conservation bodies may be able to perform this type of role. However, the same principles apply to governments, their agencies, and NGOs as they do to individuals who contract as buyers. In particular, when the PES scheme involves long-term commitment on the part of sellers to continue to supply either inputs to the provision of environmental services or the output of environmental services themselves, that length of commitment should be matched on the buyer side of the agreement. For governments and their agencies, such long-term commitments may be difficult to provide given funding cycles and/or political cycles in democracies. Any disparity between the time frame provided by the buyer and the length of supply commitment required of the sellers will result in diminished levels of trust and hence increased risk that agreements will not be fulfilled.

The prevailing legal system within which contracts are written is also of note when it comes to buyers. When PES scheme buyers are all located within one legal jurisdiction, there is no issue. However, when buyers are located across jurisdictional boundaries, the process of contracting becomes more complex, particularly with regard to enforcement. Although a PES contract may be written in one country, if a buyer is resident outside that country then enforcement is likely to be problematic. When there are multiple buyers located across multiple jurisdictions, the problems become manifest.

Contracting with governments and their agencies may also present some risks. Where a government chooses to renege on a contract, recourse to penalties may be complex if the government has control over the legal system. Without the "separation of powers," the prospect of "sovereign risk" whereby the state defaults are heightened. This is a further indication of the importance to PES scheme success of a robust underlying legal system. Similar conclusions can be drawn with regard to the presence of corruption in government and judicial circles.

The Lao PES scheme

Until the sustainable funding mechanism targeting tourists and residents is in place, seed funding for the Lao PES scheme is being provided by the World Bank (see Section 2.5). *Hence, the current buyers in the scheme are members of the international community, represented by the governments that provide funding to the operations of the World Bank. The overall buyer payment equals the amount of funding provided under the World Bank's Protected Area and Wildlife Project (PAWD). The buyers are indirectly contracted through an agreement between the following brokers: the World Bank, the Environmental Protection Fund (EPF), and the National University of Laos (NUoL) (see* Section 2.7). *They are entrusted to ensure contract compliance and penalize contract violations. In other words, the buyers have no choice but to trust the brokers that the environmental services they pay for are being provided.*

Under the recommended sustainable funding mechanism the buyer contracts would be evident through the one-off visa entry fee for tourists and the monthly electricity surcharge collected from residents. Both payment collection systems are designed to be compulsory to address free riding, which is understood to be a real concern in the context of wildlife diversity protection. Every tourist and resident would have to pay the average willingness-to-pay estimated for each buyer group (see Section 4.5). *The only way to avoid payment would be to abstain from visiting the Lao People's Democratic Republic (PDR) or using electricity, respectively.*

The contracts would not include any legal provisions that enable the buyers to enforce contract compliance. Buyers would not be able to refuse to pay if they suspected a contract violation such as the embezzlement of funds earmarked for the antipoaching patrol scheme. The buyers would have to trust that the brokers are using the funds to pay the sellers that produce the promised environmental services.

To inspire buyer confidence and build trust, it is crucial to provide information about the intended use of the payments as well as the roles, responsibilities, and accountability of the involved brokers. A comprehensive information package, potentially distributed in hard copy to tourists and to residents, may include a link to a web page containing up-to-date information on the antipoaching patrol scheme and the status of the environmental services KNOWLEDGE.

7.2 Sellers

In contrast to PES buyer contracts, seller contracts are generally more complex. This is primarily because the seller contract must specify operational aspects of the PES supply process. This is especially the case where sellers are contracted to provide inputs into the environmental services provision process rather than the contract specifying environmental service outcomes. This is because buyers will need to be confident that the outcomes they are buying will be provided through a tightly defined input process. Buyers may prefer to contract for outputs but the time frames involved may make sellers reticent to wait for outcomes to be achieved before they receive payment.

Other terms to be included in a seller contract will relate to the definition of the selling entity, amounts to be received, the timing of the payments made, conditions to be met for payment, the type of payment transfer mechanism to be used (for instance, direct payment to a bank account), the monitoring systems (to check if contractual conditions have been met), and associated penalties for breaches. In addition, there may be a need for the definition of a grievance process to be implemented when sellers become dissatisfied with buyer performance (and vice versa). Such a process would involve an identified "neutral" mediator, conditions when mediation becomes necessary, a mechanism for resolution of the identified conflict and any redress that is required beyond those associated with contract breeches.

A PES contract is an important means of ensuring that the incentives faced by sellers match the desires of the buyers. The "sticks" associated with breech penalties associated with the "carrots" associated with contract payments upon delivery constitute the incentive structure. Monitoring is key to the success of the incentive structure implemented. Payment needs to be triggered through a process of compliance testing. In brief, buyers need to be assured that payments will be made only if achievements are logged. Furthermore, the logging of achievements must be genuine and not fraudulent. Compliance tests must therefore be "costly to fake" for sellers and the compliance process must not be corrupt.

There are numerous ways of ensuring that faking a test is costly. The first line of defense against such fraud is to ensure that the test set is a genuine reflection of achievement and is hard to falsify. Technological advances can help in this regard. Cheap access to digital photography and associated locational software (including GPS tracking) can help to verify actions taken as inputs to the provision of environmental services. Penalties associated with attempts at fraud can also strengthen the incentives for compliance. In other words, sellers must believe that there is a real probability of fraud being detected by those monitoring the contract and that there is a realistic penalty associated with that fraud.

The sellers' peer group can also provide some incentives for compliance. If other, potentially competing, sellers become aware that fraud is being committed, they may assert pressure for it to stop so that the "playing field" of the competition remains fair. If the broader community in which the seller lives is dependent on the success of the contracted activity, members may also assert pressure to make sure the seller performs to contract. Sellers' may also feel some moral obligation to the rest of the community if a dependence between seller and community is established.

Of course, the same logic applying to contracts with buyers regarding the functionality of the underpinning legal and judicial system applies to seller contracts. No matter how strong a contract may be in terms of establishing incentives, if the contract cannot be enforced because of an inadequate legal system or a more broadly corrupt institutional setting, then PES schemes will face challenges. For example, if contracts are monitored and enforced by a third party—perhaps the broker responsible for establishing the PES scheme—then the third party must be immune from bribes. If that is not the case, payments made by buyers could well end up being used for the personal advantage of agents and sellers without the achievement of the production "bought" by the buyers.

The Lao PES scheme

Seller participation in the scheme is voluntary (see Section 5.1). The commitments of the participating sellers were formalized through two types of contracts: patrol contracts and community conservation agreements. The templates of the patrol contract and community conservation agreement are available in Annexes 6 and 7.
Each patrol contract defined and specified the following elements[1]:

- *Terms used in the contract*
- *Contract period*
- *Actions (inputs) to be delivered*
- *Terms of payment (including a monitoring system to ensure payment conditionality)*
- *Equipment*
- *Insurance cover*
- *Eligible recognitions*
- *Eligible other benefits*
- *Penalties for contract violations*
- *Mechanism for grievance, conflict resolution, and redress*
- *Prohibition*
- *Force majeure*
- *Limitation of liability against a third party*
- *General terms*

Each patrol contract was signed by each patrol team member and by a representative of the NUoL (on behalf of the buyers), and certified by the respective village heads to ensure community support.
Each community conservation agreement defined and specified the following elements:

- *Conservation commitments of the community (inputs)*
- *Terms of payment (including a monitoring system to ensure payment conditionality)*

---

[1] Lao service contracts were used as a template (National Assembly, 1990, Contract Law. No. 02/NA Vientiane, Lao People's Democratic Republic).

- *Eligible recognitions*
- *Penalties for agreement violations*
- *Penalties for violations of relevant regulations and laws of the Lao PDR*
- *Village Development Fund*
- *Mechanism for grievance, conflict resolution, and redress*
- *Monitoring and evaluation of the scheme*

Each community conservation agreement was signed by the respective village head (on behalf of the community) and by a representative of the NUoL (on behalf of the buyers), and certified by the respective district governor to gain support from the district authorities.

The scheme applies the "costly-to-fake" principle combined with a monitoring and penalty system designed to maximize contract compliance and avoid perverse incentives. For example, patrol teams have to provide GPS track records and take photographs of snare lines and snares before and after they are dismantled. To ensure that the whole team performed the patrol, team photographs with GPS location stamps need to be provided. Without the provision of the evidence as set out in detail in the contracts, penalties apply. The penalties range from payment reductions to the exclusion of team members and whole teams from the scheme. The large number of teams from different villages is likely to inspire peer group monitoring, which is assumed to reduce contract violations. Furthermore, the payments to the broader communities in which the team members live are dependent on patrol team performance. Similarly, violations of the community conservation agreements committed by individual community members jeopardize the payments to the broader community. In both cases, community members are likely to assert pressure to ensure compliance and hence secure payment to their respective village development funds.

Additional incentives were provided through oath ceremonies. The oath ceremonies were conducted with the assistance of monks and/or other spiritual leaders tailored to each ethnic culture represented in the respective village. Their aim was to strengthen the team members' and the community members' compliance with the patrol contract and community conservation agreement, respectively.

7.3 Brokers

The importance of brokers to the success of a PES scheme has already been emphasized: brokers can act to reduce the transaction costs that keep environmental service buyers and sellers apart. Brokers therefore play a key role in completing links between PES scheme buyers and sellers. Those links are formalized through contractual agreements between buyers, their brokers, sellers and their brokers, and potentially between the brokers for the buyers and sellers. This complexity has to be managed in a way that moderates the costs of contracting because such transaction costs can end up being just as severe as the immediate transaction costs that separate buyers and sellers when there are no brokers involved.

Brokers are able to lower transaction costs because they specialize in performing the coordination and information provision tasks involved in forming a PES "market." One reason for their ability to enjoy lower costs is economies of scale. By specializing in specific tasks and applying those skills to multiple clients, they can generate economies of scale and scope. The same principle applies if a broker can act for both buyers and sellers in a PES scheme. By acting for all parties, the broker avoids the costs of securing an agreement between brokers acting for separate parties. The contractual obligations are thus reduced to those discussed for buyers and sellers.

Failure on the part of the broker to fulfill their contractual obligations can be redressed by both buyers and sellers (the broker's principals). The importance of the operation of the rule of law is again stressed as critical. If redress cannot be achieved in the event of a breach of contract through the legal system, then the likelihood of parties fulfilling their obligations is reduced. Similarly, if government agencies take on the role of broker, they must not be above the law or otherwise corrupt.

The potential for brokers reducing transaction costs may also be enhanced by more frequent adoption of PES schemes. "Off-the-shelf" PES schemes designed on the basis of other successful schemes in similar contexts may be developed by PES brokers and offered at low cost to both buyers and sellers. Residential tenancies, incorporation documentation, and other legal agreements demonstrate the potential for lower cost deal facilitation.

The Lao PES scheme

The scheme brokers involved with contracting (World Bank, NUoL, EPF, district governors, and village heads) act on behalf of both the buyers and the sellers. This implies that they represent the sellers in the contracts with the international community (and with the tourists and the residents under the sustainably funded design of the Scheme) as well as the buyers in the contracts and agreements with the patrol teams and broader communities, respectively.

The international community (and tourists and the residents) rely on the brokers to deliver the purchased environmental service KNOWLEDGE. The patrol teams and the broader village communities rely on the brokers to pay them performing the contracted actions. While the patrol teams and the broader village communities have a mechanism for grievance, conflict resolution, and redress at their disposal if the brokers breech the contracts and agreements, such a mechanism is not available to the international community (and the tourists and the residents). Consequently, rigorous monitoring and enforcement as required as part of the World Bank's PAWD program (and through an independent third party) is crucial to ensure the scheme's success in delivering the environmental service KNOWLEDGE.

7.4 Challenges and limitations

Contracting buyers and sellers poses a range of challenges. A major challenge is to design an effective incentive structure. Such a structure usually includes both negative and positive incentives, for example, conditional payments and penalties in case of contract violations. It is crucial to strike the right balance between the "sticks" and "carrots" and customize the incentives to the local context. The costs associated with the specified penalties need to be higher than the benefits gained through cheating. An important element is the "costly-to-fake" principle. The costs to produce evidence that satisfies payment conditions needs to be higher than providing evidence gained through performing the actions specified in the contract. Otherwise, the scheme provides so-called perverse incentives. Avoiding perverse incentives is not always easy and often requires a phase of "trial and error." For example, snare manufacturing in people's backyards inspired by payments for the collection of snares would most likely result in very few snares being collected and carried out of the forest. This would reduce significantly the effectiveness of the scheme. Equally important and often challenging is to establish an effective and cost-effective monitoring system that maximizes the risk of detection. Ideally, the incentive structure provides strong positive incentives, costly penalties for contract violations, and a high risk of detection.

Another challenge is associated with embedding the contracts into a functioning legal system. The most watertight contract is futile if it cannot be enforced through the rule of law. This may be especially problematic in contexts where the legal system is corrupt, the sellers and buyers are restricted or discouraged from using the legal means available, and the accountability of brokers is questionable. In such cases an independent, trustworthy third party is of the uppermost importance to minimize contract violations and maximize their enforcement.

Communication is another challenge that may be encountered while negotiating contracts, particularly in contexts characterized by low literacy and education levels. PES scheme designers need to ensure that the sellers understand what they are signing up to. Otherwise, the effectiveness and efficiency of the scheme are likely to be severely compromised.

References and further readings

Publications of team members and collaborators of the Lao PES project:

Scheufele, G., 2016. Payments for environmental services. In: Bennett, J. (Ed.), Protecting the Environment, Privately. World Scientific Press, London.

Scheufele, G., Bennett, J., 2017. Can payments for ecosystem services schemes mimic markets? Ecosystem Services 23, 30—37.

Scheufele, G., Bennett, J., 2013. Research Report 1: Payments for Environmental Services: Concepts and Applications. Crawford School of Public Policy, The Australian National University, Canberra.

Scheufele, G., Bennett, J., Kragt, M., Renten, M., 2014. Research Report 3: Development of a 'virtual' PES Scheme for the Nam Ngum River Basin. Crawford School of Public Policy, The Australian National University, Canberra.

Scheufele, G., Smith, H., Tsechalicha, X., 2015. Research Report 4: The Legal Foundations of Payments for Environmental Services in the Lao PDR. Crawford School of Public Policy, The Australian National University, Canberra.

Tsechalicha, X., 2017. Research Report 14: Engaging Communities in a Payments for Environmental Services Scheme for the Phou Chomvoy Provincial Protected Area. Crawford School of Public Policy, Australian National University, Canberra.

Publications of authors external to the project:

Bolton, P., Dewatripont, M., 2004. Contract Theroy. MIT Press, Cambridge.

Ferraro, P.J., 2008. Asymmetric information and contract design for payments for environmental services. Ecological Economics 65, 810−821.

How did it go?

How well did the bioeconomic model predict the effectiveness of actions (inputs) in producing environmental services (outputs)? Did the buyers and sellers honor their commitments? How well did the brokers perform their roles? How well did the predictions of future demand reflect reality? Did the established "market" generate the predicted changes in social well-being? Answers to these questions provide insights into the overall Payments for Environmental Services (PES) scheme performance and offer guidance as to if and how a scheme could be modified to improve its effectiveness and efficiency. An assessment of PES scheme performance that is based around these questions should be conducted at the end of each supply period, which is formalized through the seller and buyer contract terms. The answers so produced can then be used to make any modifications to the scheme's operations before the next round of conservation auctions. Importantly, those changes would need to be embedded in the terms and conditions of the resultant supplier and buyer contracts.

The following sections outline how to assess and improve PES scheme performance. Once again, specific reference is made where possible, to the practicalities of performing this step as experienced during the Lao PES scheme. Because the Lao scheme is still in its initial phase of operation, a complete assessment is yet to be conducted. However, some lessons have already emerged that are worth noting.

The Lao PES scheme

The Lao scheme has only been in operation for about a year at the time of writing. Hence, a thorough assessment of its performance was not yet possible. What is provided here is a checklist of actions that would need to be performed to carry out a full assessment. Moreover, this should be conducted before the expiry of the current World Bank funding.

8.1 Production functions

Production functions are established on the predictions of bioeconomic models (see Section 3.1). Typically, a model of a production function is based on a range of simplifying assumptions and focuses on the biophysical system's core parameters and relationships (see Section 3.2). Were the assumptions used to model the PES scheme's input–output relationship justified? Did the model parameters capture

Buying and Selling the Environment. https://doi.org/10.1016/B978-0-12-816696-3.00008-9

all factors that were relevant in predicting the output? How well did the specified equations describe the cause—effect relationships between the parameters? How accurate were the (assumed) parameter values that were used to populate the model? And with all of that checked, the ultimate question is how accurate were the model predictions?

Of course, assessing model performance along these questions requires the collection of additional data after the scheme has been in operation for some time and particularly before a second round of contracts with buyers and sellers are negotiated and signed. Ideally, the initial PES scheme would be designed to encourage ongoing data collection, for example, through the actions performed by the sellers, so that specific additional data collection exercises are not required.

The results of this assessment can then be used to modify the model specifications and underlying assumptions. Assessing and modifying the model on a regular basis facilitate the continuous improvement of the model's predictive power. In that manner, model performance keeps improving the longer the PES scheme is operational. The transaction costs of these modifications are deemed negligible if the model is designed to be readily adjustable (see Section 3.2).

Additional data may also indicate that the actions performed by the sellers should be modified to improve their effectiveness in producing environmental services. Any modifications would have to be accounted for in the updated bioeconomic model.

The Lao PES scheme

The dynamic population model was developed to predict the population size of each of the target species. It was defined by spatial and temporal specifications as well as a range of parameters and relationships that enabled the simulation of core processes (see Section 3.1). The model was used to predict the effectiveness of the antipoaching patrols (inputs) in protecting wildlife diversity (output). In other words, the model was used to estimate production functions (see Section 3.2).

The PES scheme was designed to encourage ongoing data collection through the antipoaching patrol teams (see Section 2.2). In contrast, performing law enforcement and wildlife monitoring tasks the teams to collect data on, for example, snare line and poacher camp locations, poacher movements, poaching incidents, poaching gear, and characteristics of populations of the target wildlife species. These data can be used to check and adjust the specified assumptions, parameters, and relationships (and their respective values) as used in the models. The associated simulated core processes such as the assumed reductions in the number of poachers and snare lines because of patrols, the movement of animals and poachers, and the animal deaths from illegal poaching through snaring for each target species can also be checked and if necessary adjusted in the models.

8.2 Demand

An assessment of the demand side targets the performance of the buyers and the broker(s), in so far as they act on behalf of the buyers. Buyer performance can be assessed by checking the degree of contract compliance. The commitments listed in the buyer contracts form the basis for this assessment (see Section 7.1). If buyer participation is compulsory through the introduction of taxes or fees collected by a government authority, the performance of the broker(s) is the focus of the demand-side assessment. How effective was the payment vehicle used to collect the buyer payments? How effective was the mechanism that transfers the buyer payments to the sellers? Were there additional demand-side transaction costs borne by the broker(s) that had not been included in the calculations of the social net benefit? How large was the difference between the predicted and actual number of buyers? Are there indications that the extent of demand has changed significantly? Are there new buyer groups that should be considered in the next contract period?

The results of this assessment can then be used to modify the demand-side elements of the PES scheme such as buyer commitments, the payment vehicle, and the payment transfer mechanism. If there are strong indications that the extent of demand has changed, willingness to pay may have to be reestimated. This, however, should only be considered if the expected efficiency gains outweigh the transaction costs associated with this effort.

The Lao PES scheme

An assessment of the demand side focuses on the performances of the brokers who act on behalf of the buyers. The brokers' performances can be checked by assessing the degree to which they fulfilled their current responsibilities and commitments: Did the World Bank provide the funds as agreed? Did the Environmental Protection Fund (EPF) transfer the funds to the National University of Laos (NUoL) as agreed? Did the NUoL transfer the funds to the patrol manager responsible to make the payments to the patrol teams and the village development funds? Did the patrol manager perform her tasks as specified in her contract? Were bank accounts for the patrol teams and the village development funds established? Was the mechanism for grievances, conflict resolution, and redress established? Did patrol teams and village development funds receive the patrol equipment, insurance cover, and their payments on time and in accordance with the seller contracts? Did the patrol teams and the communities receive recognitions in accordance with their contracts? How many complaints were lodged and solved through the mechanism for grievances, conflict resolution and redress? Were there any additional transaction costs borne by the broker(s) that were not initially accounted for?

The demand assessment also needs to check and adjust (if necessary) the data used in the estimation of demand. Were the recorded tourist arrival numbers and resident growth rates consistent with the predictions made? Were there indications that the preferences of the identified buyer groups (tourists and residents) have changed? Is there evidence that new potential buyer groups have emerged? Has the Government of Lao PDR (GoL) reduced the restrictions on access to the Phou Chomvoy Provincial Protected Area (PCPPA) as to make the environmental service WATCHING available to buyers?

Additionally, the demand assessment needs to check if arrangements that ensure the sustainability of the scheme beyond the initial 3 years are in place. Has the GoL instituted an ongoing funding mechanism as recommended by the PES designers[1]? Will the EPF continue to be involved in the scheme? What future avenues for funding are available? Numerous attempts to engage mining and hydroelectricity companies operating in the Lao People's Democratic Republic were made by the PES scheme designers in the initial negotiations without success. Might they be more willing to engage with a successful scheme in place? An alternative way forward to involve these companies might require a change in legislation to facilitate compulsory offset payments.

These data collected through this assessment need to be used to check and adjust the estimations of demand (and supply in case additional transaction cost borne by the broker(s) were identified).

8.3 Supply

An assessment of the supply side looks at the performance of the sellers and the broker(s), in so far as they act on behalf of the sellers. The degree of contract compliance is the core indicator in assessing seller performance. The basis for this assessment is the commitments listed in the seller contracts (see Section 7.2). How many sellers completed their contracts? How many sellers performed the supply actions as agreed? How many sellers violated the environmental code of conduct? How many sellers received a bonus or recognition for excellence in performing their supply obligations? If any contract violations affected the amount and/or quality of inputs, the bioeconomic model would need to be updated and the production functions reestimated.

The reasons for any contract violations would need to be identified to improve compliance and, in consequence, the effectiveness and efficiency of the PES scheme. Circumstances may change or may not have been accurately predicted over the course of the initial contract period, thus causing contract breaches that were beyond the control of either seller or broker. Hence, adjusting action inputs and/or the incentive structure to reflect the changed circumstances may reduce the number of contract violations.

---

[1] However, has come to our attention that the GoL has introduced a new visa levy. However, it remains unclear if the funds go into consolidated revenue or than a specific fund earmarked for a specific purpose.

Breaches may also be caused by the poor performance of the broker(s) who represent the sellers. The basis of assessing broker performance on the supply side is the broker commitments listed in the seller contracts. How many grievances were submitted to and resolved by the mechanism for grievance, conflict resolution, and redress? Have conflicts emerged caused by grievances that were reported? Were these conflicts unresolved? What were the causes of these grievances and conflicts? Did the sellers receive the agreed payments on time? Did the sellers receive the inputs that were supposed to be provided by the broker(s)? Did the sellers receive the promised bonus payments and recognitions? Did the PES scheme managers (brokers) perform their duties as set out in the seller contracts? An assessment of seller and broker performance may have to be supported by a survey of sellers and broker(s).

The assessment of broker performance should also include the quantification of any additional supply-side transaction costs borne by the broker(s) that were not initially included in the calculations of the social net benefit.

To minimize supply disruptions, the assessments detailed earlier and all resulting contract term modifications have to be made between the end of the seller contract term and the next round of conservation auctions.

The Lao PES scheme

An assessment of the supply side focuses on the performance of the sellers and to some degree the brokers acting on their behalf. The degree of contract compliance is the core indicator in assessing the performance of the patrol teams and the broader communities. The basis for this assessment is the commitments listed in the patrol contracts and community conservation agreements. What percentage of the contracted patrols was completed? How many patrol teams and communities have earned recognitions for excellent performance? How many teams or community members have submitted a complaint through the mechanism for grievances, conflict resolution, and redress? What were the reasons for these complaints? How many teams and community members were issued a penalty? What type of penalties was issued? How many patrol team members or whole patrol teams have quit the scheme? How many team members or whole teams have violated the environmental code of conduct? What are the reasons for contract violations and team dropouts? Were there any additional transaction costs borne by the broker(s) that were not initially accounted for?

The data collected through this assessment need to be used to check if any contract violations affected the amount and/or quality of inputs. If that is the case, the bioeconomic model would need to be updated and the production functions and supply costs reestimated.

8.4 Social well-being

An overall PES scheme assessment also includes an appraisal of potential differences in social well-being estimated based on the data available at the start and the end of each supply period. That is, the benefits and costs associated with the supply of environmental services and its distribution among buyers and sellers (see Section 6.2) have to be reestimated using the additional data and assessment results. A reestimation of the social net benefits at the end of a supply period reduces the uncertainties surrounding these estimates. In other words, the ex ante benefit–cost analysis done during the PES schemes development should be verified through an ex post benefit–cost analysis. The longer the scheme is in operation, and therefore, the more reestimations are performed, the smaller these uncertainties will become.

The Lao PES scheme

The data collected through the assessment outlined earlier would need to be used to reestimate the distribution of costs and benefits across buyers, sellers, and brokers, as well as the overall social net benefit of the scheme (see Section 6.2). This allows a comparison between the predicted and actual social net benefit generated under the scheme. Closing the gap between prediction and reality reduces the uncertainties surrounding the estimates of demand and supply.

References and further readings

Publications of team members and collaborators of the Lao PES project:

Scheufele, G., Bennett, J., 2013. Research Report 1: Payments for Environmental Services: Concepts and Applications. Crawford School of Public Policy, The Australian National University, Canberra.

Scheufele, G., Bennett, J., Kragt, M., Renten, M., 2014. Research Report 3: Development of a 'virtual' PES Scheme for the Nam Ngum River Basin. Crawford School of Public Policy, The Australian National University, Canberra.

Publications of authors external to the project:

Barbier, E., Hanley, N., 2009. Pricing Nature: Cost-Benefit Analysis and Environmental Policy. Edward Elgar Publishing, Cheltenham.

Where to from here?

Payments for Environmental Services (PES) schemes, internationally, have often been touted as offering the potential to deliver a new beginning for the protection and or the restoration of environmental assets. This has been the case particularly in contexts where the chances of governments stepping in to provide and manage protected areas are slim. Such contexts extend from developed countries in which fiscal constraints hit the public environmental purse through to developing countries where the capacity of the public sector in terms of both funding and skill is lacking. However, disappointment has almost equally often been the outcome. The obstacles that PES scheme seek to overcome—the free-rider problem and associated transaction costs—are substantial. Moreover, the dangers posed by vested interests—including government agencies as well as PES scheme sellers and brokers—capturing the PES scheme process for individual financial gain are also significant. So as with most policy issues, the future prospects of PES schemes are a matter of the balance between the benefits that are available from their implementation and their costs. The goal of this chapter is therefore to review the material presented in the preceding chapters and present the major "promises" offered by PES schemes alongside the "pitfalls" likely to be encountered. Finally, an overview of the contexts in which PES schemes are more likely to succeed (and fail) is provided, along with a range of suggested actions that will improve the likelihood of success.

9.1 Promises

The fundamental promise afforded by PES schemes is the ability to generate environmental services that are in demand by the public and which in the absence of the scheme would be either not provided at all, or significantly under provided. The idea behind PES schemes is to overcome the barriers that exist between buyers and sellers of environmental services so that trade can take place in a "pseudomarket" setting.

Of specific importance in the operation of PES schemes is their capacity to allow environmental service buyers to deal more directly with environmental service sellers. The creation of a "pseudomarket" through a PES scheme, offers the prospect of more efficient provision of environmental services than what would be available either through direct governmental provision or in the absence of any external action. Without government intervention, the chances of vested interest groups

Buying and Selling the Environment. https://doi.org/10.1016/B978-0-12-816696-3.00009-0

engaging in "rent-seeking behavior" are diminished. Vested interests would not be able to lobby governments for access to their environmental service budget because they would not have one. Furthermore, without funds having to be collected and then spent by government on environmental service provision and management, the "overhead" costs of government including the costs of taxation and administration are reduced. In addition, PES schemes can ensure that supply is more responsive to the demands of buyers: only the environmental services that are in sufficient demand by buyers would be produced. Moreover, cost control is enforced on suppliers: with sellers competing against each other, buyers would seek out the best "deals" in terms of quantity and quality of supply relative to the price charged.

These are all promises that are offered through the competitive supply of goods and services in an open market. By mimicking market forces, PES schemes have the opportunity to follow suit.

However, the reasons for PES schemes being required relate to the problems faced by markets in dealing with environmental services that have public good characteristics. That implies some form of intervention. Even moving away from a PES scheme scenario in which no government intervention is involved offers promise. The free-rider problem on the demand side of the PES market may be so severe that voluntary contributions fall well short of true levels of environmental service demand. With a disciplined government acting as an agent for buyers, the supply-side strengths of a PES scheme can be engaged. Competitive suppliers responding to a brokers call for expressions of interest can ensure that cost effectiveness in supply is achieved with incentives in place to advance greater production efficiencies over time.

The operation of a PES scheme can also be used to generate greater discipline for governments. With an informed public (and a vigilant independent media backed by an open access social media) keen to see that funds collected specifically for a PES scheme—perhaps through a purpose built compulsory levy scheme—are used with good effect, democratically elected governments are more likely to remain on track to ensure delivery.

Similarly, a PES scheme that includes a process to estimate the strength of demand for the specified environmental service can discipline government spending. Enforcing a rule that restricts government funding of a PES scheme to the amount that is indicated by demand studies means that it is more difficult for vested interest groups to lobby politically sensitive governments into levels of funding that are beyond what the public want.

There is also the promise of PES schemes being founded on the voluntary contributions of environmental service buyers, even in the face of the free-rider motive. Critical in the counteraction of the free-rider motive is the concept of "crowding out." When the government is widely recognized as the provider of environmental services, members of the public have the predictable response of not donating to alternative private suppliers. Their logic is simple: Why should I pay taxes for the government to provide environmental services and then also have to pay for a private supply? In this manner, the government's actions to fund environmental services "crowd out" the emergence of private suppliers. To avoid this situation, governments can announce that their "public good" obligation to supply environmental services is limited to prespecified actions. Members of the public who have stronger demands for the environmental services than those serviced by the public sector would then need to look to private suppliers. Although they would still face the free-rider motive—they may hope that others would contribute enough to see their demands satiated without having to pay—there is also a learning effect involved. If, because of free riding, private supply does not emerge and demand remains unfulfilled, buyers would be more likely to learn that free riding has not been a successful strategy. Growing levels of private (voluntary) funding can then cause a growing recognition among the public that successful PES schemes are possible and so reduce the uncertainties surrounding the worth of making a voluntary payment. This applies to members of the public but also charitable trusts and commercial businesses wishing to fulfill some corporate social responsibility targets.

The evidence gained from implementing the Lao PES scheme demonstrates that many of these promises can be achieved. In the absence of the PES scheme, wildlife poachers were active in the border regions between Lao People's Democratic Republic (PDR) and Vietnam and the future of numerous endangered species in that area was highly uncertain. With the scheme, the level of poaching effort in the Phou Chomvoy Provincial Protected Area (PCPPA) has been reduced and at least for the immediate future, an improved level of species protection has been achieved. The people engaged in the antipoaching patrols have secured an improved standard of living because of their enhanced incomes and the villagers have expressed satisfaction with the operation of the scheme. The technical studies underpinning the PES scheme ensured that the fundamentals of the scheme were well grounded in the principles of both ecology and economics. These are all positive achievements and demonstrate what can be done in the somewhat difficult context provided by the Lao PDR setting. In the past, this context has provided major obstacles for intended PES schemes.

The Lao PES scheme experience should not be viewed as either infallible or entrenched. Its intention was to provide a demonstration of what could be achieved. As such, its current scale provides only limited security for the endangered species it is seeking to protect. A high security level would only come from a geographical expansion of the current scheme. Nor is its future secured beyond the tenure of the World Bank funding that supports it on the demand side. The scheme does however perform a valuable role in demonstrating some of the pitfalls that can confront PES scheme implementation and operation. These are outlined in the next section.

9.2 Pitfalls

The failure of PES schemes to emerge over the last 2 decades as the preeminent modus operandi for environmental protection and restoration indicates that the promises set out in the previous section are not easily won. The fact that normal markets have failed to emerge for the environmental services involved is indicative of the extent of transaction costs keeping buyers and sellers apart. Designing a PES scheme that can effectively reduce these transaction costs to the point where buyers and sellers find it worthwhile to engage in exchange is therefore no easy task.

The efficiency of a PES scheme is related to the extent and quality of the information gathered and utilized in the process of its design. Information is not free. Markets work to generate information through the self-motivated business of exchange. Pseudomarkets such as PES schemes require that information to be gathered through brokers who may not be profit motivated. Without that motivation, the incentive for efficient data gathering is diluted. Ignoring the lack of motivation, there may also be a lack of expertise among those who would be logically expected to take on a brokering role. For instance, government departments in developing countries are unlikely to have the requisite skills within their work force. Even research agencies and universities may not be able to meet the requirements in the context of a poor nation. Furthermore, the high frequency of staff turnover in many agencies means that training initiatives that are intended to develop the required skills among the staff only have short-term payoffs. The long-term sustainability of PES schemes is thus compromised as the "human capital" formed through the initial design and implementation stages moves on to other opportunities (that are not necessarily PES related).

The staff capacity issue is exacerbated by the sometimes long "gestation period" of a PES scheme. Particularly when a "greenfield" site is being considered for a PES scheme, it can take a considerable period of time to carry out the necessary implementation steps. For instance, the conservation auction process including the necessary community consultation phases may take several months. Over that time, the staff responsible have a higher chance of "moving on." However, it is also possible that agencies, impatient to see some PES "action" on the ground will seek to dilute the rigor of the underpinning analysis and implement a scheme that is at risk of creating inefficiencies.

Potentially more serious if governments are involved, the broking role may be captured by rent seekers aiming to maximize their personal gains. The risk is that brokers can circumvent the information gathering process and implement a scheme that is highly inefficient. For instance, very little information is required to use funds dedicated to a PES scheme for a simple income redistribution scheme that involves local people being paid money in return for no/little environmental restoration/protection services. Similarly, a political decision to dedicate part of a government's consolidated revenue to pay environmental service suppliers may not be based on the extent of demand but rather on the political expediency of votes originating among those being paid. A local bureaucracy may also seek to capture rents from a PES scheme to enhance its size and prestige.

The efficiency gains achieved from gathering and using supply and demand information to establish a pseudomarket have to be traded-off against the costs associated with the information so collected.

No doubt, the biggest stumbling block for any PES scheme remains the raising of funds contrary to the free-rider motive. Although voluntarism has some potential in this realm, securing PES scheme sources remains a challenge. The result often involves funds sourced from government or international agency sources. This in itself presents some pitfalls. First, vested interests can capture the government agencies involved for personal advancement. But in addition, publicly sourced funds can be fickle. For governments, decisions to allocate funds are notoriously inconsistent over time, especially when governments change hands. Different governments have different priorities and renewal of contracts may not always proceed successfully. Similarly, international agencies experience changing priorities. PES scheme sellers can therefore be left without a market without much warning even when significant investments in both time and money have been sunk. These concerns relate to both democracies and authoritarian regimes.

The complexities of setting up a PES scheme can also lead to it being relatively inflexible. Given the time and effort required to "get it right," the temptation for those involved in implementing a PES scheme is to be reluctant to move away from the initial parameters of the scheme—particularly the price paid/received. This is a potential pitfall because it ignores the constantly changing context that underpins any market, pseudo or otherwise. Supply and demand conditions are likely to be changing frequently and considerably. Ignoring those changes could leave a PES scheme irrelevant and without buyers or sellers because the price set initially is no longer appropriate. The smaller the scheme, the more dramatic such changes might be for the future of a PES scheme. On a small scale, flexibility to match changing "market" conditions is likely to be more difficult to achieve. For instance, if a supplier drops out of the market—perhaps because they experience an increase in their opportunity costs—then finding a replacement supplier may be problematic.

A related pitfall for PES schemes is found in the capacity of the underpinning legal system to provide an adequate and secure framework for the establishment and enforcement of contractual agreements. Part of this potential problem for PES schemes relates to property law. If the resources providing the environmental services in question (land, water, vegetation, etc.) do not have secure and well-defined property rights assigned to them, then confusion regarding responsibilities and rights arise. In brief, if the rule of law is not in place, there are added uncertainties to be overcome by both buyers and sellers. These add to the transaction costs of exchange and provide a further barrier to trade. Even if the rule of law is in place, PES schemes will still encounter problems if enforcement of the law (through the police force and/or the courts) is either corrupt or expensive.

The Lao PES scheme experience illustrates these pitfalls. For instance, the search for funding for the scheme went on for 3 years and finally relied on the establishment of a loan to the Lao government from the World Bank. The limited time period for this loan coupled with a requirement for the funds to be repaid implies serious issues once the initial contracts for supply have expired. Although the Lao Government has implemented a visa surcharge as advised by the project team, the channeling of the funds so raised to the ongoing financing of the scheme is far from secured.

The costs of the research effort underpinning the design and implementation of the scheme were also significant. Although they were born externally from Laos and were not only targeted at the Lao context (with the results having broader significance and application), they do illustrate the costs of achieving the complexity of the pseudomarket so established. These costs can now be considered "sunk" and the information so gleaned is available for wider use at far lower marginal cost.

The operation of a PES scheme in a highly regulated socialist nation also added to the transaction costs, particularly in terms of the extent and complexity of the nation's bureaucracy and legal system. Part of the problem in Lao PDR relates to the "ownership" of property rights. The legal system in the socialist state deems the land resource as the property of the state. That means the sellers in the PES scheme do not hold ownership rights to the land or the resources therein (notably the wildlife and forest assets). The uncertainty of incentive that this situation creates is a further contributor to transaction costs.

9.3 A way forward

PES schemes do have potential. However, that potential needs to be carefully nurtured if it is to be realized.

First, it is important to ensure that any PES scheme follows fundamental market-based processes and principles. Allowing breaches of those principles beyond justifiable and transparent compromises to address practicalities of the context opens the door to rent seeking and inefficiencies. The price paid to sellers should be determined through the interaction between marginal costs and marginal benefits. The process of contracting with sellers needs to be open, transparent, and competitive. The extent of funding must be guided by information on the strength of demand for the environmental services involved. In addition, the (ecological) productivity relationship between contracted inputs and demanded outputs needs to be understood so that demand and supply can be linked.

Second, the rule of law needs to be invoked as strongly as possible to provide a secure underpinning for the PES scheme. Importantly, the legal principle of precedent—that finding from past cases provide guidance as to what behaviors are unlawful—means that PES schemes will not be continually subject to legal challenge. Once established and tested, PES schemes under a jurisdiction are likely to run without significant ongoing legal expense.

Similarly, securing highly visible political support for a PES scheme is important so that the potential for rent seeking/corruption in the bureaucracy and among politicians is limited. This is harder to achieve in some jurisdictions than in others. Being able to call on the support of a PES scheme "champion" within the relevant political jurisdiction is also an advantage, especially when legislative or bureaucratic compliance is required.

Third, drawing on supply, demand and productivity information already gathered and analyzed from other existing and successful PES schemes can be an important strategy to lower the transaction costs of setting up a new scheme. The good news here is that as more PES schemes are developed, more "transferrable" information becomes available and the marginal transaction costs fall.

A related point concerns the available expertise of people engaged in the PES scheme process. As more schemes are enacted, more people are becoming trained in the specifics of PES scheme information elements. More people are capable of organizing conservation auctions among potential sellers. More people are skilled in conducting environmental demand revealing surveys. More people can effectively model environmental production functions to link supply actions to demand. Again, this effectively reduced PES scheme transaction costs.

A key aim of this volume has been to provide guidance to those contemplating a PES scheme. This is part of the transaction cost reduction process. Throughout this volume, the experience of establishing a scheme has been used to illustrate what can and might work. Contexts are different across different environments and political jurisdictions. Hence, not all the lessons set out may be pertinent. However, following the general rules of aiming for efficiency and sharing of surplus among buyers and sellers while moderating the costs will go a long way in guiding actions.

References and further readings

Publications of team members and collaborators of the Lao PES project:

Bennett, J., Kyophilavong, P., Scheufele, G., 2017. Research Report 18: Key Findings and Policy Recommendations on Payments for Environmental Services Schemes in Lao PDR. Crawford School of Public Policy, Australian National University, Canberra.

Scheufele, G., 2016. Payments for environmental services. In: Bennett, J. (Ed.), Protecting the Environment, Privately. World Scientific Press, London.

Scheufele, G., Bennett, J., 2017. Can payments for ecosystem services schemes mimic markets? Ecosystem Services 23, 30—37.

Publications of authors external to the project:

Barbier, E., Hanley, N., 2009. Pricing Nature: Cost-Benefit Analysis and Environmental Policy. Edward Elgar Publishing, Cheltenham.

Coggan, A., Whitten, S.M., Bennett, J., 2010. Influences of transaction costs in environmental policy. Ecological Economics 69, 1777—1784.

Frank, R., 2003. Microeconomics and Behavior. McGraw-Hill, Boston.

McCann, L., Colby, B., Easter, K.W., Kasterine, A., Kuperan, K.V., 2005. Transaction cost measurement for evaluating environmental policies. Ecological Economics 52, 527—542.

Mankiw, N.G., 1998. Principles of Economics. The Dryden Press, Harcourt Brace College Publishers, Fort Worth.

Tietenberg, T., Lewis, L., 2009. Environmental and Natural Resource Economics. Pearson International Edition, Boston.

Expert survey

Expert survey to assist the development and implementation of a PES scheme in the annamite mountain range

Background

One of the pilot Payments for Environmental Services (PES) schemes that are being developed within the project "Effective Implementation of PES in Lao PDR" will focus on wildlife conservation in the Annamite Mountain Range. The potential areas to be targeted by conservation actions include the Nam Chat/Nam Pan PPA, the Nam Chouan PPA, and the Phou Chomoy PPA located at the border with Vietnam. We understand that the main problem to endangered wildlife in these areas is poaching through snare lines. Consequently, the suggested focus will be on the destruction of snare lines and the removal of snares.

We understand that the snare lines consist of a "fence" made of sticks and other material found in the forests with gaps where the actual snare (a wire) is installed. Antipoaching patrols could reduce the negative impact of poaching if they destroy the snare lines and collect the snare wires. Based on initial expert interviews and information on the experiences gained in the Nakai-Nam Theun National Protected Area, we have developed a first draft of an antipoaching scheme:

(1) Each area to be divided into 1 km^2 grid cells using 1:50,000 topographic maps; each cell marked by a center point (GPS reference).

(2) Core zones" of high conservation priority will be identified—these will be subject to the most intense patrolling.

(3) Patrol teams consist of four villagers and two government officials. Patrol teams may camp together to feel safer.

(4) Each patrol team is instructed and expected to visit the center points of a specified series of assigned grid cells during each patrol.

(5) Assignment of cells is such that the effort is not predictable to poachers. However, the allocation must be sensible—allocated cells should be adjacent to allow an efficient route.

(6) Patrols have to complete a certain number of cells per day of patrolling: four cells per day in the dry season and two cells in the wet season. For example, if patrols were 10 days long, they would have to complete 20 cells in the wet

141

season and 40 cells in the dry season. If they are faster that is ok as long as they can provide evidence that they completed all cells.

(7) Evidence must be provided to prove that the patrol has visited the specified cells: recorded GPS "waypoints" and track logs.

(8) This evidence reduces the possibility of patrols actively steering the patrolling away from poaching activities and provides accurate records of patrol effort.

(9) To be effective, each cell should be visited once per month and all snare lines and snares should be destroyed and monitoring activities undertaken. Patrols have to destroy the snare lines and the snare themselves need to be collected. The snares themselves, photographic and GPS evidence would ensure that "snares" cannot be "made at home."

(10) Patrol teams would be encouraged to take photographs of specified endangered species and to record tracks and droppings to collect occupancy data to help better understand species distributions and thus better target future patrolling.

(11) Payments will be made per day of patrolling (but no more than the specified days per cells to be visited) and additional payments will be made (a) per photo of targeted endangered species, (b) per collected snares. Later, we may also include payments for leeches, droppings, etc.

(12) GIS software will be used to manage the spatial patrol data obtained by GPS, including timing and locations of patrols, snares, wildlife encounters, and poacher encounters. The free software SMART might be used to manage the conservation data.

This survey has been sent to a wide range of people carrying different knowledge sets. Please ignore those questions that you think are out of your area of expertise.

Questions part A: species

We realize that accurate estimates for many of the parameters are unlikely to be known. We are asking for your best guesses based on experience.

Please use as much space as you need for your answers.

1) What do you think are the key endangered, rare, or threatened species in the target areas (Nam Chat/Nam Pan and Phou Chomvoy PPAs) that are significantly affected by snares?

2) Do you think that all these species equally affected by snares, or are some more vulnerable than others?

3) Which species should be the focus of the antipoaching scheme?

4) Are there "core zones" of high priority which you would recommend to be subject to the most intense patrolling?

Questions part B: proposed antipoaching scheme

We realize that accurate estimates for many of the parameters are unlikely to be known. We are asking for your best guesses based on experience.

Please use as much space as you need for your answers.

5) Does the proposed scheme and design make sense to you?

6) Do you think the proposed scheme would be effective in improving the conservation of key wildlife species?

7) Do you disagree with any of the points?

8) Do you have suggestions how to improve the scheme?

9) What do you think villagers, on average, would need to be paid per day of patrolling to engage in the proposed antipoaching scheme?

10) How much would we have to pay per collected snare to encourage dismantling the snare lines and collecting the snares during patrolling?

11) Photographs of which species should be rewarded?

12) How much would we have to pay per photograph of specified species to encourage patrol teams to take photographs?

13) Would the payment per photograph have to be different for different species?

Questions part C: snares

We realize that accurate estimates for many of the parameters are unlikely to be known. We are asking for your best guesses based on experience.

Please use as much space as you need for your answers.

14) From your experience, what is the typical length of a snare line?

15) What is the range of snare line lengths poacher use, that is, what do you think is the shortest and the longest snare line poacher use?

16) How many snares would typically be set in one snare line?

17) What is the typical spacing of snares along a line?

18) Are snares ever deployed individually, or always along snare lines with multiple snares?

19) From your experience, what is the density of snares maintained by poachers in the target areas?

20) How often do you think snare lines are checked by poachers, and captured animals removed and snares reset?

21) If a snare line is dismantled by a patrol (fence dismantled and snare wires removed) how long do you think it would be until poachers replaced it?

Questions part D: patrols

We realize that accurate estimates for many of the parameters are unlikely to be known. We are asking for your best guesses based on experience.

Please use as much space as you need for your answers.

22) Do you think a minimum number of days of patrolling in a row is required (e.g., 5 or 10 days) to make patrolling feasible and/or effective? If so, how many days?

(13) If a patrol visits a 1 km^2 grid cell that has a snare line in it, what is the probability that they will find the snare line? Please remember that each patrol team is instructed and expected to visit the center points of a specified series of assigned grid cells during each patrol.

23) If you think the probability is low, how could the chance of finding the snare line be increased?

24) If a patrol finds a snare line, will they then surely find all the snares along the line?

25) Do you think that having more knowledge about the distribution of the key species would allow patrols to be better targeted, thus leading to better conservation outcomes?

Questions part E: species affected by threats other than snares

We realize that accurate estimates for many of the parameters are unlikely to be known. We are asking for your best guesses based on experience.

Please use as much space as you need for your answers.

26) Are there endangered, rare, or threatened **animal species that are not significantly affected by snares,** but still significantly affected by poaching (e.g., by collecting or shooting)?

27) Are there endangered, rare, or threatened **trees or plants** that are significantly affected by illegal logging or collection?

28) Would regular patrolling be likely to reduce the shooting, collecting, or logging of these species (by deterring poachers for example)?

29) What percentage reduction (if any) in poaching/collecting/logging (in percentage or number of animals per species) do you think a monthly patrol would cause? What about a fortnightly patrol? Or a weekly patrol?

Thank you for participating in this survey! Your assistance is much appreciated.

If you have any questions regarding this survey or our project please do not hesitate to contact us.

Choice modeling questionnaire[1]

Main survey—tourists

Language: English.

> *Mode: Personal interviews with "paper" choice cards submitted in envelope.*
> *Attention: Do not write on show cards.*

Hi, my name is *[interviewer's name]*. I am looking for international tourists visiting Laos to be part of a survey about managing wildlife in the Phou Chomvoy Provincial Protected Area.

Q1 **Are you visiting Laos as a tourist?**
   Y   Yes
   N   No *[End survey by saying "OK. Thanks for your time"]*
Q2 **Would you be willing to give me 20 min of your time to answer some questions?**
   Y   Yes
   N   No *[End survey by saying "OK ... Thanks for your time"]*

Thank you.

The survey is being done by the National University of Laos together with the Australian National University and the University of Western Australia.

Participants in this survey will remain anonymous. Your name would not be recorded, and your answers are strictly confidential.

If you have any questions about this survey, please contact the project coordinator.

Hand over the contact information card.

Q3 **Did you have to get a tourist visa to come to Laos?**
   Y   Yes
   N   No
Q4 **Where are you from?**
   1 Australia
   2 Austria

---

[1]The show cards are available on request from the lead author.

3	Belgium
4	Brunei
5	**Cambodia**
6	Canada
7	**China**
8	Denmark
9	Finland
10	France
11	Germany
12	Greece
13	India
14	Indonesia
15	Israel
16	Italy
17	Japan
18	Korea
19	Malaysia
20	**Myanmar**
21	Netherlands
22	New Zealand
23	Norway
24	Philippines
25	Russia
26	Singapore
27	Spain
28	Sweden
29	Switzerland
30	Taiwan
31	**Thailand**
32	United Kingdom
33	USA
34	**Vietnam**
35	Other Africa and Middle East
36	Other Americas
37	Other Asia and Pacific
38	Other Europe

*If respondent is from one of the following countries, **end survey** by saying "OK … Thanks for your time but we are not surveying visitors from your country."*

5	**Cambodia**
7	**China**
20	**Myanmar**
31	**Thailand**
34	**Vietnam**

*If respondent is from any other country, **continue**.*
Let me now give you some background on what this survey is about.
The survey consists of three parts:

- First I will tell you about the Phou Chomvoy Provincial Protected Area and the current state of the wildlife there.
- Second, I will show you some different options to manage wildlife in the Protected Area and will ask you which option you like best.
- I will finish with some questions about yourself.

Part 1

Give the show card bundle to the respondent, with Show Card 1 "Location of the Phou Chomvoy Provincial Protected Area" visible.
Please have a look at this map. *[Show Card 1 "Location of the Protected Area"]*
The Phou Chomvoy Provincial Protected Area is in the Northern Annamite Ranges in Bolikhamxay Province on the border with Vietnam. The Protected Area is about 22,300 ha.

Q5 Have you heard of the Phou Chomvoy Provincial Protected Area?
   Y   Yes
   N   No *[Skip Q6 and Q7]*
Q6 Have you visited the Phou Chomvoy Provincial Protected Area?
   Y   Yes
   N   No

   Please have a look at the next card. *[Show Card 2 "Condition of the Phou Chomvoy Provincial Protected Area"]*

Q7 What do you think is the current condition of the Phou Chomvoy Provincial Protected Area? Please rate the condition on a scale from 1 to 5, where 1 is very good, 2 is good, 3 is fair, 4 is bad, and 5 is very bad.
   1 Very good
   2 Good
   3 Fair
   4 Bad
   5 Very bad
   6 I do not know

   Please have a look at the next card. *[Show Card 3 "Background"]*
   This card gives you some background information about the Phou Chomvoy Provincial Protected Area. Please read the information.
   Please have a look at the next card. *[Show Card 4 "Examples of threatened wildlife"]*

These are examples of species that are still found in the Protected Area but are threatened with extinction because of poaching.

Please have a look at the next card. *[Show Card 5 "Typical village in the region around the Protected Area"]*

This is a photograph of a typical village in the region around the Protected Area.

If respondent has a tourist visa (Q3: Yes) proceed with Statement 1.
Statement 1:

Patrols to reduce poaching, payments to village funds to improve local people's living conditions, and setting-up and maintaining basic tourist facilities would cost money. This would be collected through a tourist levy on top of the visa fee.

*If the respondent **does not** have a tourist visa (Q3: No) proceed with Statement 2.*
Statement 2:

Patrols to reduce poaching, payments to village funds to improve local people's living conditions, and setting up basic facilities for tourists would cost money. This would be collected through a tourist levy paid at international border checkpoints.

Continue with the survey for both types of respondents.

This is necessary because the Lao government does not have enough money for these new management actions.

An independent agency would make sure that the money collected through the tourist levy would be spent only on antipoaching patrols, improving the living conditions of local people and setting-up and maintaining basic tourist facilities.

Part 2

In this survey, we want to find out how you would like wildlife in the Phou Chomvoy Provincial Protected Area to be managed. I will ask you to make some choices between different future management options.

Each option has different combinations of management actions and would therefore have different outcomes.

Please have a look at the next card and read the explanations about possible outcomes. *[Show Card 6 "Outcomes"]*

Please have a look at the next card. *[Show Card 7 "Choice question example"]*

Here is an example of the choices I am going to ask you to make. Please have a look at this sample choice question.

The respondent does not have to make a choice. It is an example to explain to the respondent how making choices works.

Explain to the respondent using Show Card 7 what the outcomes would be if they choose Option 1. Repeat the explanation for Option 2.

If you choose **Option 1** by ticking the box "my choice" this means as follows:

- There would be no antipoaching patrols.
- None of the threatened species would get additional protection.

- Current poaching levels would not be reduced. Poaching levels would stay at 25%.
- Tourists would not be able to visit the Protected Area.
- There would be no payments to village funds.
- You would not have to pay a tourist levy.

If you choose **Option 2** by ticking the box "my choice" this means as follows:

- There would be antipoaching patrols.
- Patrols would protect 20 different species.
- Patrols would reduce poaching of the 20 protected species. Poaching levels would fall to 20%.
- Tourists would be able to visit the Protected Area.
- 500 households would benefit from improved living conditions through payments to village funds.
- You would have to pay a tourist levy of 10 US dollars per visit to Laos.

I am now going to ask you to make 5 choices like that. I would like you to choose your preferred option in each question.

The option that receives the greatest support will be put in place. This means all international tourists would have to make the payment required for that option.

When making your choices please consider the following:

- You would have to pay a tourist levy every time you visit Laos; and
- Your available income is limited and you have other expenses.

There are no right or wrong answers. We are interested in what you think.

Your answers are important. The results of this survey will be used to decide how wildlife in the Phou Chomvoy Provincial Protected Area will be managed in the future.

I would now like you to make your choices.

Record your and the respondents ID on the back of the booklet and give the *choice card booklet and an envelope to the respondent.*

This booklet gives you 5 choices to make.

Once you have made your choices, please put the booklet into the envelope and give it back to me.

Once the respondent has returned the envelope, continue with the questionnaire.

Thank you for making your choices.

Q8 **When making your choices, did you always choose Option 1, the "no new management actions" option that did not require any payment?**
   Y   Yes
   N   No *[Proceed to Q10]*

Please have a look at the next card. *[Show Card 8 "Reason"]*

Which of the following statements most closely describes the main reason why you always chose Option 1, the "no new management actions" option that did not require any payment?

Please tell me which category applies to you.

Q9   *Record category of main reason*

Record <u>one category number only</u>—the one that most <u>closely describes the main reason.</u>

Part 3

We are now in the final part of the survey. I would like to finish by asking you some questions about your background.

We ask these questions to make sure that we have a representative sample of tourists visiting Laos.

Let me stress again that your answers are strictly confidential and that all information will be kept anonymous.

Q10   *Record gender of respondent*
  M   Male
  F   Female
Q11   **Are you traveling alone?**
  Y   Yes *[Proceed to Q14]*
  N   No
Q12   **How many adults and children (under 18 years) are traveling in your travel party, including yourself?**

  Record number

Q13   **Who are you traveling with?** *[Record all categories that apply.]*
  1 Spouse/partner/boyfriend/girlfriend
  2 Child/children
  3 Other family members
  4 Friend(s)
  5 Other
Q14   **How many days did you stay in Laos?**

  Record number of days

Q15   **How much money did you spend during your stay in Laos including prebooked hotels and transportation within Laos?**

  *Record amount **and currency***

Q16   **What is the highest level of education you have obtained?**
    1 Primary education
    2 Secondary education (e.g., high school, college, year 10, year 12, trade certificate, trade diploma)
    3 Tertiary education (e.g., university degree, undergraduate, postgraduate, graduate diploma; doctorate, PhD)

Please have a look at the next card. *[Show Card 9 "Age"]*

Q17   **This card lists age categories. Please tell me which category applies to you.**

Record category.
My last two questions are about your income.
Please have a look at the next card. *[Show Card 10 "Household income"]*
This card lists categories of **gross household income** earned last year (wages/salaries, government benefits, pensions, allowances, and other income). Please tell me which category most closely applies to **your household**. Please do not deduct tax, superannuation contribution, health insurance, amounts salary scarified, or any other automatic deductions.

Q18   *Record **gross household income** category*

Thank you. Please have a look at the next card. *[Show Card 11 "Individual income"]*
This card lists categories of **gross individual income** earned last year (wages/salaries, government benefits, pensions, allowances, and other income). Please tell me which category most closely applies to **you**. Please do not deduct tax, superannuation contribution, health insurance, amounts salary scarified, or any other automatic deductions.

Q19   *Record **gross individual income** category*

Thank you for participating in this survey. Have a nice flight.

Information booklet for villagers*

Payments for environmental services scheme Bolikhamxay Province
Information booklet for communities

Effective Implementation of Payments for Environmental Services in Lao People's Democratic Republic (PDR) 2015

http://ipesl.crawford.anu.edu.au

*All illustrations were created by Phonethip Keomala.

Topic 1: What are Payments for Environmental Services (PES) schemes and how do they work?

ผอกเริาผู้ใจในกานปักปักธักสาสัดป่า   ผอกเริาต่อๆกานจ่ายเพื่ออะบุลักสัดป่า

Our PES scheme has two goals:

* Protecting wildlife; and
* Making a positive contribution to the livelihoods of communities.

We want to achieve this by paying people to protect wildlife in the Phou Chomvoy Provincial Protected Area (PCPPA).

Topic 2: Who can be part of the PES scheme?

All communities that are around the PCPPA are invited to be part of our PES scheme.

* Being part of our PES scheme is voluntary.
* Being part of our PES scheme means that your community would agree to a range of commitments to protect wildlife in return for payments and other benefits.
* The commitments and how your community will use the payments would be negotiated with your community through the development of a Community Action Plan.

Community Action Plan:

* Recognition of boundaries of the PCPPA

- Recognition of legal restrictions on use of wildlife within the PCPPA
- Agreement on conservation commitments to protect wildlife within the PCPPA
- Support of patrolling scheme within the PCPPA
- Development of rules for the use and management of the community payments
- Establishment of a mechanism of grievance, conflict resolution, and redress at the village level

Support in monitoring and evaluating social impacts of the PES scheme.

Topic 3: What will the antipoaching patrol scheme look like?

Villagers will form patrol teams that consist of five villagers, of which two or three are members of the village militia. These teams will perform regular patrols within the PCPPA.

A "patrol" is defined as a number of days of patrolling during which the team has to visit the preset series of grid cells. The series of grid cells to be visited and the patrol pattern will be assigned by the patrol manager and will change every time. The required number of grid cells to be visited will vary by seasons and by differences in the difficulty of terrain.

Topic 4: What will your community get?

Patrol teams will get payments, health, and accident insurance to cover their patrolling activities and field clothing including boots and hats.

Patrol teams who fulfilled their obligations and made exemplarily efforts over the course of 1 year will be recognized as "trusted wildlife guardians."

Your community will get payments for honoring the commitments to wildlife protection.

These payments will include a fixed amount and a variable amount (equal to a percentage of the payments to the patrol teams of your community) into a village fund.

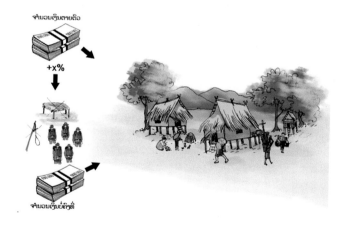

If your community has fully honored their commitments over the course of 1 year your village will be recognized as a "trusted wildlife guardian."

Topic 5: How will the price per patrol be determined?

In our PES scheme we will use an approach called bidding that is similar to purchase orders.

Concept of bidding:

- For a range of prices, potential patrol teams are asked to "bid" the number of patrols per year they would like to do.
- The "bids" are sealed. The teams only know their own bid but not that of any other teams.
- Once we have the bids we will match them with the amount that buyers want to pay. This gives us the price per patrol.
- The price will be set so that the benefits are shared between buyers and sellers.
- Bidding is a self-selecting process. Each team decides how many patrols they would like to do at this price.

Topic 6: How will the money get to your community?

An advance of 25% on the first regular payment will be paid to the patrol team immediately after signing the Patrol Contract. The remaining 75% payment will be made after the first 3 months. Subsequent payments to the patrol teams will be made every 3 months.

The payments to the patrol teams will transfer directly into patrol team bank accounts held by a district bank.

Patrol teams will be able to withdraw money from the team's bank account through a check. All team members have to sign it to withdraw money.

The payments to the community will be transferred directly into the bank account of the Village Development Fund held by a district bank.

- The Village Development Fund will be managed according to the principles developed by the LuxDev project.
- Withdrawal from the bank account will follow the procedures of the Village Development Fund.

Topic 7: What will happen if your community does not honor the commitments?

Community members who do not honor the community commitments will be penalized. The penalties will be developed in consultation with the Village Development Committee during the development of the Conservation Agreement.

Patrol teams and patrol team members who do not honor their commitments will be penalized.

- If a patrol team does not conduct the patrol or does not provide the full set of evidence, they will not get paid for that patrol.

- The team payment per patrol is the sum of every team members' portion. Team members who do not complete the patrol do not get their portion. That is, the team payment will be reduced.

- If a patrol team does not complete the number of patrols they signed up within a 3-month period, the three-monthly payment will be reduced by 20%.

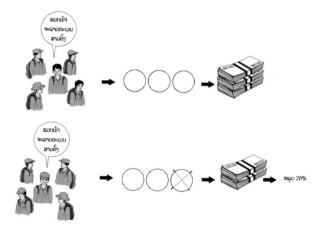

Topic 8: How can your community file a complaint?

Community members who have a complaint about the scheme can try to solve the issue directly with the patrol manager. However, community members who want to file a complaint will be able to use the mechanism for grievance, conflict resolution, and redress without having talked to the patrol manager first.

The use of this mechanism is free of charge.

This mechanism will protect the rights and interest of all PES participants or people affected by the PES scheme: community members, patrol teams, and the patrol team manager.

Topic 9: How will we formalize the commitments?

The community commitments will be formalized through a Conservation Agreement between your community and PONRE.

- The Conservation Agreement will be signed by the Village Head, the Village Development Committee, the Village Forestry Unit, DONRE, PONRE, and the Office of the District Governor.
- An "oath ceremony" will be held with the assistance of monks and/or spiritual leaders tailored to each ethnic culture represented in the village.

The patrol team commitments will be formalized through a Patrol Contract.

- Potential patrol teams will be briefed and consulted on the Patrol Contract as well as the performance and payment conditions before signing. Teams can withdraw before signing if they do not want to be involved.
- The Patrol Contracts will be signed by PONRE and all team members and endorsed by the Village Head, the Village Development Committee, and the Village Forestry Unit, DONRE, and the District Governor's Office.
- Oath ceremonies" will be held with the assistance of monks and/or spiritual leaders tailored to each ethnic culture represented in the village.

Topic 10: How can your community register an expression of interest?

If your community is interested in being part of the PES scheme, the Village Head/ Village Development Committee are invited to register an expression of interest (community engagement).

- The Village Head/Village Development Committee need to fill out the form "Expression of interest (community engagement)" and submit it in a box labeled "Expression of Interest."
- "Expression of interest (community engagement)" forms will be given to the Village Head/Village Development Committee.
- The Village Head/Village Development Committee can submit an expression of interest up to 2 weeks following the completion of this consultation workshop.
- An expression of interest is not an obligation to participate in the PES scheme. It just tells us that your community as a whole is interested in being part of the PES scheme.

Villagers who are interested in being part of a patrol are invited to register an expression of interest (patrolling scheme). To register, villagers need to form groups of five people, of which two or three are members of the village militia.

- Each group needs to fill out an "Expression of interest (patrolling scheme)" form and submit it in the box labeled "Expression of Interest."
- "Expression of Interest (patrolling scheme)" forms will be available at the end of the consultations.
- Groups can submit their expressions of interest up to 2 weeks following the completion of this consultation workshop.
- An expression of interest is not an obligation to participate in the patrol scheme. It just tells us that you are interested in participating in the bidding for patrolling and would like to participate in a training workshop.
- Note: Villagers of your community can only become part of the patrolling scheme if the Village Head/Village Development Committee have registered community interest.

Topic 11: What will be the next steps?

Incorporation of community feedback
(October 2015 – December 2015)

Development of Community Action Plan and Conservation Agreement
(December 2015 – January 2016)

Training workshops and bidding for patrolling
(December 2015 – January 2016)

Evaluation of bidding results
(February 2016)

Signing of Conservation Agreements and Patrol contracts
(March 2016)

Contact Details
Project Coordinator/Lao PDR
Dr Phouphet Kyophilavong
Vice-Dean and Associate Professor
Faculty of Economics and Business Management
National University of Laos

Vientiane Capital, Lao PDR
T +856 21 770067
E Phouphetkyophilavong@gmail.com

Project Leader/Australia
Dr. Jeff Bennett
Professor
Crawford School of Public Policy
The Australian National University
Canberra, ACT, 0200, Australia
T +61 2 6125 0154
E jeff.bennett@anu.edu.au

Project Leader/Lao PDR
Mr. Saysamone Photisat
Deputy Director General
Ministry of Natural Resources and Environment
Department of Forestry Resource Management
Vientiane Capital, Lao PDR
T +856 21 261187
E saiphothisat@yahoo.com
Project website: http://ipesl.crawford.anu.edu.au
Illustrations: Mr. Phonethip Keomala

Conservation auction training manual[1]

Bidding for patrol contracts in Bolikhamxay Province
Manual for facilitators

This training manual contains eight topics and includes a power point presentation that compiles photographs, illustrations, and examples to help explaining the more complex topics. Instructions on how to use these materials are provided within in each module. The workshop will be conducted within 1 day.

Topics

- Recap of suggested antipoaching patrol scheme
- Environmental code of conduct
- Physical and cultural resources chance-find procedures
- Patrol contracts
- Bidding training
- Bidding for patrol contracts

Patrol scheme design

Show illustration of a patrol team (Slide 1)

- Let us recap how our antipoaching patrol scheme will work.
- Villagers will form patrol teams that consist of five villagers, of which two or three are members of the village militia.
- These teams will perform regular patrols within the PCPPA.
- Patrol teams will be properly trained by experts on patrol techniques and use of tools necessary to record field data, such as GPS, maps, and patrol data form.
- One member of each patrol team will act as patrol team leader. Patrol team leaders will be identified during the patrol team training.
- The PCPPA will be treated as one area. That means any team may patrol in any part of the PCPPA.

---

[1]The slides are available upon request from the lead author.

- The location, timing, and tasks of the patrol teams will be planned and managed by a patrol manager.
- The patrol manager will send a half-yearly report to the Village Forestry Units of all participating villages.

Patrol tasks

- Let us recap what teams will have to do while they are patrolling.
- Patrol teams have two overall core tasks:
  - protecting wildlife, and
  - monitoring wildlife.
- Let us first have a look at the wildlife protection actions.
- Patrol teams will dismantle snare lines and collect snare wires during patrolling. Snare wires will have to carry the snare wires back to the village. Patrol teams will have to provide evidence to show that they have removed the snare lines (e.g., photographs of existing and dismantled snare line with location, date, and time stamps; collected snare wires).
- Patrol teams will dismantle illegal camps during patrolling. Patrol teams will have to provide evidence to show that the camp has been dismantled (photographs of existing and dismantled camp with location, date, and time stamps).
- Patrol teams will record poaching incidents and record evidence.
- Patrol teams will report poachers they encounter by recording conversations, asking for signed statements; and alerting Provincial Office for Natural Resources and Environment (PONRE); Provincial Office for Forestry Inspection (POFI).
- Patrol teams will confiscate gear used for poaching wildlife. Evidence that need to be presented: confiscated gear, confiscation form with signatures of patrol teams and poachers, and photos with date, time, and location stamp. Patrol teams will hand over poaching gear to the patrol manager (snare wires) or the police (for example guns).
- Patrol teams will confiscate animals and animal parts, including dead animals/parts, and release animals that are alive. Patrol teams need to provide evidence (photographs of scenes with location, date, and time stamps).
- Patrol teams will issue warnings to local poachers using warning forms to record relevant information.
- Patrol teams will arrest poachers who are not Lao citizen.
- Let us now have a look at the wildlife monitoring tasks of a patrol team.
- Patrol teams will record any direct sightings of key wildlife species using a data form (information on species, number of individuals, GPS location, date, and time).
- Patrol teams will record encountered signs (e.g., tracks, scat) of key wildlife species every 300 m (information on species, GPS location, and time) using a data form.

Patrol team management

Show map divided into grid cells marked with center points and patrol pattern (Slide 2).

- Let us have a closer look at how the patrol teams will be managed.
- The PCPPA will be divided into 1 km^2 grid cells using maps; each cell will be marked by a center point using a device that allows us to fix a location on the map and in the field. This devise is called a GPS.
- Each patrol team will visit a series of grid cells over 7 days by following a specified pattern (e.g., visiting the center points, following a zigzag line). That is, a "patrol" is defined as 7 days of patrolling during which the team has to visit the preset series of grid cells.
- The series of grid cells and the patrol pattern will be assigned by the patrol manager and will change every time.
- The required number of grid cells will vary by seasons (less during the wet season than during the dry season) and by differences in the difficulty of terrain (less in densely vegetated and steeper areas than on flatter and less densely vegetated areas).

Show photograph of a group photo with stamps (Slide 3).

- Patrol teams will have to provide evidence that the whole patrol team has visited the assigned grid cells in the specified pattern (recorded GPS coordinates; photographs of team members with date, time, and location stamps of the patrol team at the start, middle, and end of each patrol day).
- Patrol teams have to act in accordance with the environmental code of conduct, which will be discussed later today.

Equipment for the patrol teams

Show photographs of equipment (Slide 4), name the items, and explain their purpose.

- Camera within-built GPS function (remains the property of NUoL)
- GPS unit (remains the property of NUoL)
- SIM cards for mobile phone or radios (remains the property of the NUoL)
- Maps (remain the property of NUoL)
- Compass (remains the property of NUoL)
- Binoculars (remain the property of NUoL)
- Paper notebook and record sheets
- Backpack (remains the property of NUoL)
- Flysheet (remains the property of NUoL)
- Hammock (remains the property of NUoL)
- First aid kit (remains the property of NUoL)
- Mosquito repellent

- Antileech socks
- Hat
- Boots
- Field clothing (same color for all patrol teams; marked as "village patrol")

Discussion

- Do you have any questions?
- Do you have comments?
- Do have any concerns?

Environmental code of conduct

- Patrol teams have to comply with the environmental code of conduct.
- The environmental code of conduct will ensure that patrolling in the PCPPA will damage neither wildlife nor their habitat.
- Let us have a look what this code will include:
  - Patrol teams are not allowed to hunt any wildlife for food during an anti-poaching patrol.
  - Patrol teams will minimize disturbance of wildlife.
  - Patrol teams will not make camp in ecologically fragile areas and will minimize the cutting of vegetation and site clearing.
  - Patrol teams will dismantle their camps and putting off their cooking fire completely before continuing the patrol.
  - Cigarette stubs must be extinguished completely.
  - Patrol teams will carry out any garbage (including cigarette stubs) and not discard it within the PCPPA.
  - Patrol teams are expected to be a role model for the wider community. Breaching the Community Conservation Agreement by any member of the Patrol Team during antipoaching patrolling or off-duty will be treated as a violation of this Environmental Code of Conduct.

Show core massages of the environmental code of conduct (Slide 5).

Discussion

- Do you have any questions?
- Do you have comments?
- Do you have any suggestions?
- Do have any concerns?

Physical and cultural resources chance-find procedures

- Physical and cultural resources (PCR) are defined as movable or immovable objects, sites, structures, groups of structures, and natural features and landscapes that have archaeological, paleontological, historical, architectural, religious, aesthetic, or other cultural significance.

- Chance-find procedures have been developed to mitigate against damage or loss to PCR. The following chance-find procedures are relevant for the patrol teams:
  - Patrol teams will not move or interfere with a suspected Chance-Find.
  - Patrol teams will immediately report a suspected Chance-Find to the Village Head and a representative of the District Office of Natural Resources and Environment.

Show core massages of the physical and cultural resources chance-find procedure (Slide 6).

Discussion

- Do you have any questions?
- Do you have comments?
- Do have any concerns?

Patrol contracts

Show the illustration of signing a Patrol Contract (Slide 7).

- Let us have a look what a patrol contract will look like.
- Each patrol team will sign a Patrol Contract with the District Governor(s) Office who represents the buyers. That is, the "agreement" between each team and District Governor(s) will be made legally binding.
- Teams can withdraw before signing if they do not want to be involved.
- The Patrol Contracts will be signed by all team members and endorsed by the Village Head, the Village Development Committee, and the Village Forestry Unit, DONRE and the District Governor(s).
- Oath ceremonies" will be held with the assistance of monks and/or spiritual leaders tailored to each ethnic culture represented in the village.

Show patrol contract template (electronic pdf document).

- Let us now have a look what the Patrol Contract will look like.

Discussion

- Do you have any questions?
- Do you have comments?
- Do have any concerns?

Overview

- Everyone here is part of a team of five people who has submitted a valid interest in participating in the bidding for a patrol contract.
- Before we will conduct the actual bidding later today, we will do some bidding training. This training will give you some experience how bidding works so that you can make decisions that will benefit your team. We will use examples other than patrolling to get you engaged into bidding.
- Once we have completed the training, all groups will have the opportunity to participate in the actual bidding for patrol contracts.
- Participation in the bidding for patrol contracts is voluntary. Having completed the training does not force you to participate.
- However, if you do not participate in the bidding, you would not have a chance to participate in the patrol scheme.

Example 1 (berberine vine collection)

- Let us start with our first example.
- Buyers are interested in purchasing an NTFP, berberine vine.
- A berberine vine contract would have the following conditions:
  - The berberine vine needs to be of adequate quality.
  - The suppliers have to deliver the berberine vine to the buyer's shop at their own costs.
- Potential berberine vine suppliers are asked to "bid" the amount of berberine vine they would like to collect for a given range of prices.

  Show bid table for berberine vine (Slide 8).

- How much berberine vine would you want to sell at each price?
- Suppose, three people of your village would be interested in selling berberine vine. To be able to sell berberine vine, they need to participate in the bidding.
- In the bidding, they need to state how much berberine vine they each want to sell at each price.
- People want to sell berberine vine if the price they get at least covers their costs of supplying it. Otherwise, they make a loss.
- So the first step in the bidding process is to estimate how much it costs you to supply berberine vine.
- These costs include the following:
  - Time to collect berberine vine. Let us consider the cost of time. While collecting berberine vine, you will not have the time to do other work. The more you use your time on collecting berberine vine, the less time you will have for other work or jobs. If this work is important, or if the jobs pay wages, this means you will lose these benefits if you spend more time on collecting berberine vine. For example, spending 2 days on collecting berberine vine may only cause low costs (you may have some spare time). Yet, spending

more and more days on collecting berberine vine might mean that you have to cancel a job that pays a wage or you have to leave work on your farm undone. Your costs get higher and higher.
- Transportation to the place where you collect berberine vine.
- Transportation of the berberine vine to the buyer's shop.
- Can you think of any other costs?
- Do you think these costs are the same for all three suppliers? Why not?
- Once each supplier has estimated their costs, they decide how much berberine vine they each want to supply at each price.
- It is important that suppliers do not discuss their bids with each other. Each supplier just knows his or her own bid but not that of any other bidder.

Show bid table for berberine vine (Slide 9)

- At a price of Kip 2500, Mr. Kham would like to collect and sell 10 kg. At a price of Kip 3500, he stated that he would like to collect and sell 30 kg. At a price of Kip 4500 he would like to collect and sell 40 kg.
- Once all suppliers have submitted their bids, they will be matched with the amount that buyers want to pay. This gives us the price per kilogram of berberine vine.
- That is, the price will be set so that the benefits are shared between buyers and sellers. It is a win—win situation. Let us have a look at the example.
- Let us assume that the set price is Kip 3500. In our example, Mr. Kham stated in his bid that he would like to collect and sell 30 kg, and Ms. Chanh stated that she would like to collect and sell 20 kg. The buyer will purchase 30 kg from Mr. Kham and 20 kg from Ms. Chanh.
- As you can see, it is a self-selecting process. Everyone decided how much berberine vine they want to collect and sell at this price.
- Now, you might think why not bid at a price that is higher than the costs of collecting berberine vine? Let us have a look at the bid of Mr. Boonma. He stated in his bid that he would not like to collect and sell any berberine vine until the price is at least Kip 4500. As the set price is only Kip 3500 he would not be able to sell any berberine vine. If Mr. Boonma had started his bidding at a price that was equal to his costs, he would have been able to sell some.
- Let us have a look at Mr. Kham. As his costs to collect 10 kg of berberine vine are about Kip 2500, he starts bidding at that price. As a result, he has a competitive advantage and will be able to sell more berberine vine than Ms Chanh and Mr. Boonma. If, on the other hand, he had only bid at prices that are higher than his costs, he would have lost his competitive advantage and might not have been able to sell any berberine vine.

Example 2 (collecting bamboo shoots)
- Let us now do a bidding exercise in your teams.
- Let us now assume that suppliers would work in teams of five people.

Ask all bidding teams to sit together as teams.

- A buyer is looking for suppliers of bamboo shoots around the year.
- A bamboo shoot contract would have the following conditions:
  - The bamboo shoots need to be of adequate quality.
  - The buyers will come to your village to collect the bamboo shoots from suppliers.
- Suppose your team is interested in bidding to supply bamboo shoots.
- The bidding form shows the following prices per kg of bamboo shoots.

Show bidding form "bamboo shoots" (Slide 10).
Give each team the work sheet "costs."

- Please estimate in your team the costs of your team to supply bamboo shoots. Consider the following costs:
  - Your time to collect bamboo shoots. How much time would you need to collect a certain amount of bamboo shoots? What else could you do with your time? Instead of collecting bamboo shoots, you could spent the time doing other work or earning income from other jobs. The costs of time may increase the more time you spent on collecting bamboo shoots. For example, spending a few hours per week on collecting bamboo shoots may only cause low costs (you may have some spare time). Yet, producing more and more bamboo shoots might mean that you have to cancel a job that pays a wage or that you have to leave jobs on your farm undone. Your costs of collecting bamboo shoots will get higher and higher.
  - Costs of transport to get to the areas where bamboo shoots grow and return home.
  - Food to take while collecting bamboo shoots. However, remember, you would also have to eat if you did not collect bamboo shoots.
  - Can you think of any other costs?
- Once your team has estimated their costs, you need to decide how many kg of bamboo shoots your team wants to supply at each price.
- It is important that your team do not discuss their bids with other teams. Each team just knows their own bid but not that of any other teams.

Give each team the work sheet "bidding for rattan cane."

- Please record in the work sheet "bidding for bamboo shoots" how many kg of bamboo shoots your team wants to supply for each price. Completed worksheets will be collected by a project team member.

Include the bids of all teams into the table "bidding for bamboo shoots" and show to teams (Slide 11).

- Now let us assume that the determined price paid per kg of bamboo shoots is LAK 7000.

Explain results.

Example 3 (construction work)

Ask all biddings teams to sit together as teams.

- Let us do one more bidding exercise in your teams.
- A buyer is looking for suppliers to do some construction work as casual/hired labor.
- Let us assume that suppliers would work in teams of five people.
- A construction work contract would have the following conditions:
  - The team has do work the number of days they have signed up for.
  - The buyer will allocate tasks that need to be completed per day to get paid.
  - The work will be available within a time period of 12 months.
- Suppose your team is interested in bidding to do construction work.
- The bidding form shows the following prices per day. As you can see, there are two different price ranges. One for a time period during which you are busy with planting and harvesting rice ("busy" season) and for a time period that is less busy ("quiet" season).

Show bidding form "construction work" (Slide 12).
Give each team the work sheet "costs."

- Please estimate in your team the costs of your team to do construction work. Consider the following costs:
  - Your time to do construction work. Instead of doing construction work, you could spent the time doing other work or earning income from other jobs. The costs of time may increase the more time you spent doing construction work. For example, spending a few days per month doing construction work may only cause low costs (you may have some spare time). Yet, spending more and more days doing construction work might mean that you have to cancel a job that pays a wage or you have to leave jobs on your farm undone. Your costs of doing construction work will get higher and higher.
  - Costs of transport to get to the construction side.
  - Food to take while doing construction work. However, remember, you would also have to eat if you did not do construction work.
  - Can you think of any other costs?
- Once your team has estimated their costs, you need to decide how many days of construction work your team wants to bid at each price.
- It is important that your team do not discuss their bids with other teams. Each team just knows their own bid but not that of any other teams.

Give each team the work sheet "bidding for construction work."

- Please record in the work sheet "bidding" how much construction work your team wants to do for each price. Completed worksheets will be collected by a project team member.

Include the bids of all teams into the table "bidding for construction work" and show to teams (Slide 13).

- Now let us assume that the determined team prices paid per day is LAK 250,000 and LAK 300,000.

Explain results.

Next steps

- Now let us talk about what happens next:
  - Bidding for patrol contracts: In a few minutes, all teams will have the opportunity to bid for patrol contracts. Please remember that bidding for patrol contracts is voluntary. Having completed the training does not force you to make a bid. However, teams who do not participate in the bidding would not have a chance to participate in the patrol scheme.
  - Evaluation of bids: The project team will then evaluate the bidding results of all eight villages, announce the price per patrol, and offer contracts to successful bidding teams.
  - Oath ceremony: All patrol team members will participate in the patrol team oath ceremony.
  - Signing of Patrol Contracts: The patrol teams and the relevant GoL agencies will sign the Patrol Contracts.
  - Training of patrol teams and start of patrolling: Once the patrol team training has been completed, patrolling will commence.

Requirements to participate in the bidding for patrol contracts

- Before we start with the bidding, let us check that you all fulfill the following requirements to bid:
  - All bidding teams have submitted a valid "register of interest" form and thus consist of five members, of whom two or three members are village militia.
  - Team can participate in the bidding even though not all team members are present. However, the members who are present need to have the authority to represent the whole team.
  - Only team members who completed the bidding training are allowed to participate in the bidding.
  - All team members (or their representatives) understand the tasks, responsibilities, obligations, and benefits of participating in the patrol scheme as stated in the patrol contract, which we had a look through together earlier today.
  - Bids will only become valid if all team members have signed the bidding forms before the project team's departure.

Bidding form

Show the bidding forms "patrolling" (slide 14).

- To submit a team bid, each bidding team will have to fill out two bidding forms: One for patrolling during the "busy" season, and one for patrolling during the "quiet" season.
- Let us recap what we mean by "patrol":
  - A "patrol" is defined as 7 consecutive days of patrolling during which your team has to visit the preset series of grid cells. Your team may complete the patrol 1 day early or 1 day late as long as all assigned grid cells have been visited.
  - The number of grid cells your patrol team has to visit will vary by seasons (less during the wet season than during the dry season) and by differences in the difficulty of terrain (less in densely vegetated and steeper areas than on flatter and less densely vegetated areas).
- Let us have a look what we mean by "number of patrols":
  - The number of patrols refers to the number the team is willing to perform during the "busy" season and during the "quiet" season within 1 year for the next 3 years.
- Let us have a look what we mean by "team price per patrol":
  - The "team price per patrol" refers to a price for 7 days of consecutive patrolling done by a team of five people of whom two or three members are village militia.
  - If you want to know how much that is per day and per individual team member, you can have a look at the conversion table.
  - Example: If the "team price per patrol" is LAK 1,400,000, the team price per day is LAK 200,000 and the team member price per day is LAK 40,000. However, remember, you cannot bid for individual days and as individual team members. You have to bid as a team of five people for patrols, which are each 7 days of consecutive patrolling.

Bidding team tasks

- Let us have a look what each bidding team has to do before you submit your bid.

  1. Calculate the costs of your team of doing patrols as defined in the patrol contract. Calculate these costs separately for the "busy" season and the "quiet" season.

  - Costs may include food consumed during patrolling, and your time. The cost of your time should be estimated from the perspective of other work you could do or income you could earn if you did not do patrolling. Do not include the costs of transportation to and from the PCPPA. You will be paid

a fixed average allowance to cover your travel costs to and from the boundary of the PCPPA.
- Your team should make sure that the estimated costs are as close as possible to your real situation.

2. In the Tendering Form, write down the number of patrols per year that your team would do at each proposed price. Do this for the "busy" season and the "quiet" season separately.
- Base your decision on the number of patrols your team is willing to do on your estimated costs. However, you also need to think of the benefits you will receive in addition to the payments per patrol: (1) payment per collected snare (LAK 2000); (2) payment per dismantled poaching camp (LAK 10,000); (3) health and accident insurance; (4) recognition in supporting your family and community through wildlife protection work.
- So you might think, why not bid at a price that is higher than the team's costs of patrolling? If your costs of patrolling are low and you bid at lower prices, your team will get more patrols. If, on the other hand, your team only bids at prices that are higher than your costs, you lose your competitive advantage and might only get few or no patrols.
- The number of patrols you state for each price should be agreed to unanimously within your team.
- Do not let other teams see your team's estimated costs and bids. You want to be competitive! Remember, you compete with all teams of all eight villages for patrols. The more expensive your team is, the less patrols you will get.

3. In the Tendering Form, write down your team members' names. Each team member needs to sign.

4. Put your team's completed Tender Form into the provided envelope, seal it, and give it to the PES project team.
- Once we have the bids of all teams from all eight villages, we will match them with the amount that buyers want to pay. This gives us the "team price per patrol" for the "busy" season and the "team price per patrol" for "quiet" season.
- We then have a look how many patrol each team has said they would like to do at "team price per patrol" for the "busy" season and the "team price per patrol" for "quiet" season.
- Each team will be notified about the number of patrols your team get for the "busy" season and for the "quiet" season for each of the next 3 years, which will be included in your Contract.

Final questions

- Do you have any final questions or concerns before we start with the bidding?

Give each team a piece of paper to estimate their costs, one bidding form and one envelope.

Supervise the bidding process. Make sure that teams do not talk to each other. Assist teams by answering question and by explaining the process if required. **BUT, do not fill out the form for them!**

Community action plan

Lao People's Democratic Republic
Peace Independence Democracy Unity Prosperity

Province: Bolikhamxay

District: [District name]

Village: [Village name]

Date:

On the protection of biodiversity in the Phou Chomvoy provincial protected area

- Pursuant to the Forestry Law No. 06/NA, dated December 24, 2007.
- Pursuant to the Law on Wildlife and Aquatic Animals No. 07/NA, dated December 24, 2007.
- Pursuant to the Decree on Protected Areas No. 134/Government, dated May 13, 2015.
- Pursuant to the Agreement of the Vice Prime Minister, Minister of Finance, Vice President of the Environment Protection Fund's Board of Management regarding the Approval of the EPF Subproject funding supported by the World Bank No. 008/BoM.EPF, dated July 8, 2016.
- Pursuant to the Agreement of the Governor of Bolikhamxay Province No. 516/ BXP dated July 14, 2017 on Permission for the Faculty of Economics and Business Management, National University of Laos to Conduct a Research Project on Implementation of Payments for Environmental Services in the Phou Chomvoy Protected Area in Bolikhamxay Province.

The Community Action Plan outlines the actions required to implement a Payments for Environmental Services (PES) Scheme. The PES scheme involves eight villages and aims to protect biodiversity within the Phou Chomvoy Provincial Protected Area (PCPPA).

The Community Action Plan was jointly developed by the villagers and the project "Effective implementation of payments for environmental services in Lao PDR" during community consultations in 2015 and 2016, with the agreement and acknowledgment of all relevant government authorities at the village, district, provincial, and national level.

The Community Action Plan is formalized through a Community Conservation Agreement.

Action 1: biodiversity protection within the Phou Chomvoy provincial protected area

1.1 All villagers will familiarize themselves with the boundaries of the PCPPA.

1.2 All villagers will familiarize themselves with the location of wildlife core zones within the PCPPA (which are recognized by locals despite no demarcations on the ground).

1.3 All villagers will interpret these wildlife core zones as Total Protected Zones until the formal subzoning of the PCPPA into Total Protection Zone and Controlled Use Zone has been completed.

1.4 All villagers will comply with the following restrictions on the use of wildlife and forest resources within the PCPPA:
- Buying and selling of any wildlife is prohibited.
- Hunting of endangered wildlife species mentioned in Categories I and II of the Wildlife and Aquatics Law is prohibited.
- The use of unsustainable hunting gear and methods including guns, rifles, explosives, chemicals, poison, electricity, and snares is prohibited.
- Hunting wildlife during the animals' gestation periods is prohibited.
- Hunting wildlife with offspring is prohibited.
- Hunting wildlife within the Total Protection Zone is prohibited.
- Logging is prohibited.
- Land clearing for agricultural and other purposes is prohibited.

1.5 Villagers will control collectively any violations of the Community Conservation Agreement. Villagers agree to inform the Patrol Manager, the Village Development Committee, the Village Head, and authorities at the district level of any violation of the agreement to sanction any violators.

1.6 The Patrol Manager will provide the Village Head with an appropriate number of copies of the up-to-date list of wildlife categories referred to in the Wildlife and Aquatic Law and other legal documents relevant to protected area management.

1.7 The Village Head will make the list of wildlife categories referred to in the Wildlife and Aquatic Law and other legal documents relevant to protected area management accessible to all villagers free of charge.

1.8 The Village Head and the Village Development Committee will support the antipoaching patrol scheme within the PCPPA.

Action 2: Establishment of a village development fund

2.1 Your community will adopt the principles developed by the Luxembourg Agency for Development Cooperation to manage and audit the Village Development Fund:

- The Village Development Fund can be used as a grant scheme and a credit scheme. Your community will decide if you want to have a grant scheme only or a mixture of both.
- The credit scheme is a village-owned microfinance system from which members can take loans. However, as poor and vulnerable households may be unable or reluctant to take loans, at least 20% percentage of the payments to the Village Development Funds must be available for the grant activities that benefit poor and vulnerable households whose livelihoods might be negatively affected through the biodiversity protection actions.
- The grant scheme will be managed by the Village Development Committee. The objective of this scheme is to assist the communities in making small-scale, one-time investments in activities that improve the living standards and quality of life of the population. Men and women from all households will be involved by democratic voting in to select activities.
- Activities that cannot be funded include the following:
    - × New settlements or expansion of existing settlements outside the area defined by the Participatory Land Use Planning or in any zone not gazetted for agriculture or habitation inside the PCPPA.
    - × Activities that lead to damage or loss of cultural property, including sites having archeological (prehistoric), paleontological, historical, religious, cultural, and unique natural values.
    - × Construction of new roads, road rehabilitation, road surfacing, or track upgrading of any kind inside the PCPPA and in general any construction expected to lead to negative environmental impacts.
    - × Introduction of nonnative species, unless these are already present in the vicinity or known from similar settings to be noninvasive, and introduction of genetically modified plant varieties.

× Forestry operations, including logging, harvesting, or processing of timber and nontimber forest products; however, support to sustainable harvesting and processing of nontimber products inside the Controlled Use Zone is allowed if accompanied with a management plan for the sustainable use of the resources.

× Forestry operations on land or in watersheds in a manner that is likely to contribute to increased vulnerability of your community to natural disasters.

× Conversion or actions causing degradation of natural habitat and any unsustainable exploitation of natural resources including nontimber products.

× Production or trade in wildlife products or other products or activity deemed illegal under Lao PDR laws, regulations, or international conventions and agreements, or subject to international bans.

2.2 Your community will support the training of the managers of the Village Development Fund that will enable them to perform their tasks.

Action 3: Establishment of a mechanism for grievance, conflict resolution, and redress at the village level

3.1 Your community will establish a Village Mediation Unit that comprised a selected group of villagers, including ethnic minorities, women, and representatives of other vulnerable groups in the village. The Village Mediation Unit will be put in charge of managing the grievance process at the village level as follows:

■ Grievances will be reported to the Village Mediation Unit.

■ The Village Mediation Unit will document the grievance. The document will be signed/fingerprinted by the grievant for processing.

■ All grievances will be recorded in the Village Grievance Logbook, which will be kept by the Village Mediation Unit. The Village Grievance Logbook will be accessible to all villagers at all times free of charge.

■ Within 5 days after receipt of the grievance, the Village Mediation Unit will meet the Complainant to discuss (mediate) the grievance and will advise the complainant of the outcome.

■ Any investigation at the village level will be completed within 14 days of receipt of the grievance.

■ If the Complainant is satisfied with the outcome, the issue will be closed. The Complainant will provide a signature as acknowledgment of the decision.

- If the Complainant is not satisfied with the outcome or an investigation and action/compensation at a higher level will be required, the issue will be transferred within 1 month to the District Grievance Steering Committee for further action.
- All villagers will have the additional option of raising concerns through the participatory monitoring and evaluation process and seek for resolutions at the annual district level meeting (Action 4).

3.2 Your community will ensure that the members of the Village Mediation Unit will participate in a training workshop that will enable them to perform the tasks involved with managing the mechanism for grievance, conflict resolution, and redress.

3.3 The Village Mediation Unit will inform all villagers about their right to submit grievances to the mechanism of grievance, conflict resolution, and redress.

3.4 The Village Mediation Unit will explain to all villagers how the mechanism of grievance, conflict resolution, and redress works.

3.5 The Village Mediation Unit will adequately address concerns of all villagers arising through the PES scheme in a prompt and timely manner.

3.6 Your community will offer access to the Village Mediation Unit to all community members free of charge, given that the project has provided support to village development (through contribution made to the Village Development Fund).

Action 4: Support of a participatory process that will monitor and evaluate the social impacts of the PES scheme

4.1 The Village Development Committee will organize quarterly meetings at the village level. During these meetings, all villagers will have the opportunity to discuss concerns, suggest improvements, and report any negative impact that has occurred but not yet resolved. The Village Development Committee representatives will pay particular attention to ensure that vulnerable people and minority ethnic groups have the opportunity to participate and voice their opinions and concerns during these meetings. Representatives of the Village Mediation Unit will participate in these meetings at the village level to discuss outstanding grievances or issues that have been raised.

4.2 Two representatives of your community will attend an annual meeting at the district level. One representative will be a member of the Village Development Committee, the other a villager nominated by your community. If the Village Development Committee representative is a male, then the second village representative should be a female, or the vice versa. The representatives will be encouraged to share their community's perspectives on the performance of the PES scheme, suggestions for improvement, outstanding grievances, and other relevant issues. Measures to improve project performance and resolve outstanding grievances will be also discussed and agreed on during the annual meetings at the district level. The meeting organizers will take minutes. Progress on actions agreed at the meeting will be discussed in the meeting to be held in the following year.

Action 5: Implementation of physical and cultural resources "chance-find" procedures

5.1 Physical and Cultural Resources are defined as movable or immovable objects, sites, structures, groups of structures, and natural features and landscapes that have archeological, paleontological, historical, architectural, religious, aesthetic, or other cultural significance. Your community will implement the following chance-find procedures to mitigate against damage or loss to Physical and Cultural Resources:

- Patrol teams will report immediately any suspected Physical and Cultural Resources find to the village head and a representative of the District Agricultural and Forestry Office.
- A suspected Physical and Cultural Resources find will not be moved or interfered with.
- All work potentially impacting on the find should be suspended while the village head and the District Agricultural and Forestry Office representative assess the find.
- The District Agricultural and Forestry Office representative and the village head will immediately mark the location of the Physical and Cultural Resources and, with assistance of the Patrol Manager, take necessary precautions to protect the site from further disturbance, including limiting access to the site.
- If the find contains suspected human remains, the District Agricultural and Forestry Office representative and the village head will be required to notify the relevant District Administration immediately and take instructions from the District Administration.

- The District Agricultural and Forestry Office representative and the village head will need to record the depth of the artifact if buried and document, with photographs, the artifact in situ.
- The District Agricultural and Forestry Office representative and the village head will need to prepare a chance-find report.
- The chance-find report must be submitted to the Provincial Department of Information, Culture and Tourism and the Provincial Agricultural and Forestry Office within 48 h.

Signatories:

Village Head

Faculty of Economics and Business Management, National University of Laos
Certify by:
The Head of District Governor's Office

Community conservation agreement

Lao People's Democratic Republic

Peace Independence Democracy Unity Prosperity

Province: Bolikhamxay

District: [District name]

Village: [Village name]

No.:   /

Date:

On the protection of biodiversity in the phou chomvoy provincial protected area

- Pursuant to the Forestry Law No. 06/NA, dated December 24, 2007.
- Pursuant to the Law on Wildlife and Aquatic Animals No. 07/NA, dated December 24, 2007.
- Pursuant to the Decree on Protected Areas No. 134/Government, dated May 13, 2015.
- Pursuant to the Agreement of the Vice Prime Minister, Minister of Finance, Vice President of the Environment Protection Fund's Board of Management regarding the Approval of the EPF Subproject funding supported by the World Bank No. 008/BoM.EPF, dated July 8, 2016.
- Pursuant to the Agreement of the Governor of Bolikhamxay Province No. 516/BXP dated 14 July 2017 on Permission for the Faculty of Economics and Business Management, National University of Laos to Conduct a Research Project on Implementation of Payments for Environmental Services in the Phou Chomvoy Protected Area in Bolikhamxay Province.

The Community Conservation Agreement formalizes the Community Action Plan that outlines the actions required to implement a Payments for Environmental

189

Services (PES) Scheme. The PES scheme involves eight villages and aims to protect biodiversity within the Phou Chomvoy Provincial Protected Area (PCPPA).

This Community Conservation Agreement is made and entered into by [*insert village name*], represented by the Village Head and the Faculty of Economics and Business Management, National University of Laos representing the PES scheme.

The content of the Community Conservation Agreement was jointly developed by the village [village name] and the project "Effective implementation of payments for environmental services in Lao PDR" during community consultations in 2015 and 2016, with the agreement and acknowledgment of all relevant government authorities at the village, district, provincial and national level.

An oath ceremony will be conducted with the assistance of monks and/or spiritual leaders tailored to each ethnic culture represented in the village to support this agreement. The oath will contain the following words: "We the villagers of [*insert village name*] swear that from now on we will abide by the Community Conservation Agreement, will not take actions that are damaging to biodiversity in the PCPPA and will actively support the implementation of an antipoaching patrol scheme."

This Community Conservation Agreement may be modified if any of its conditions are not suitable to the situation. All parties that have signed the present agreement will be informed of and will acknowledge and agree on all changes and modifications done.

In consideration of these premises, the Village and the Faculty of Economics and Business Management, National University of Laos, intending to be legally bound, do hereby agree as follows:

Article 1: Biodiversity protection within the phou chomvoy provincial protected area

1.1 All villagers agree to familiarize themselves with the boundaries of the PCPPA.

1.2 All villagers agree to familiarize themselves with the location of wildlife core zones within the PCPPA (which are recognized by locals despite no demarcations on the ground).

1.3 All villagers agree that these wildlife core zones will be interpreted as Total Protection Zones until the formal subzoning of the PCPPA into Total Protection Zone and Controlled Use Zone has been completed.

1.4 All villagers agree to comply with the following restrictions on the use of wildlife and forest resources within the PCPPA:
   - Buying and selling of any wildlife is prohibited.
   - Hunting of endangered species mentioned in Categories I and II of the Wildlife and Aquatic Law is prohibited.

- The use of unsustainable hunting methods and gear including guns, rifles, explosives, chemicals, poison, electricity, and snares is prohibited.
- Hunting wildlife during the animals' gestation periods is prohibited.
- Hunting wildlife with offspring is prohibited.
- Hunting within the Total Protection Zone is prohibited.
- Logging is prohibited.
- Land clearing for agricultural and other purposes is prohibited.

1.5 Villagers agree to control collectively any violations of the Community Conservation Agreement. Villagers agree to inform the Patrol Manager, the Village Development Committee, the Village Head, and authorities at the district level of any violation of the agreement to sanction any violators according to Article 3.

1.6 The Patrol Manager agrees to provide the Village Head with an appropriate number of copies of the up-to-date list of wildlife categories referred to in the Wildlife and Aquatic Law and other legal documents relevant to protected area management.

1.7 The Village Head agrees to make the list of wildlife categories referred to in the Wildlife and Aquatic Law and other legal documents relevant to protected area management accessible to all villagers free of charge.

1.8 The Village Head and the Village Development Committee agree to support the antipoaching patrol scheme within the PCPPA. The scheme will engage villagers as biodiversity protection agents. It will provide an alternative source of income to villagers who might be affected by potential livelihood losses through biodiversity protection actions and to villagers with limited livelihood opportunities.

1.9 The Village Head and the Village Development Committee agree to implement Physical and Cultural Resources "chance-find" procedures to mitigate against damage or loss of these resources through patrolling as specified in the Community Action Plan.

Article 2: benefits

2.1 The village receives payments to support biodiversity protection within the PCPPA. The payments will be made over 3 years and consist of a fixed amount and a variable amount. The fixed amount of LAK [*insert total amount*] will be paid annually. The variable amount will be equal to 5% of the payments to the patrol teams of your community. The Patrol Manager will calculate the variable payments every 3 months on the performance of the patrol teams of your community and requests payment. The Faculty of Economics and Business Management at the National University of Laos will pay the requested amount to the village and agrees to send regular payment reports to the Environmental Protection Fund and the Provincial Agricultural and Forestry Office for monitoring.

2.2 The Village Head, the Village Development Committee, and the Faculty of Economics and Business Management at the National University of Laos agree that the fixed and variable payments to the village stipulated in Article 2/Section 2.1 will be transferred directly into the account of the Village Development Fund (Article 5) held by a district bank.

2.3 The village will be recognized as a "Trusted Biodiversity Guardian" if all villagers have fully honored their commitments over the course of 1 year. The village will receive a certificate presented by a high ranking government official.

Article 3: penalties for noncompliance

3.1 Any household of the village not complying with the restrictions on the use of wildlife and forest resources set out in Article 1, Section 1.4 of this agreement will be penalized as follows:

- First transgression: The community member's household will be given a verbal warning by the Village Head in the presence of the Patrol Manager.
- Second transgression: The community member's household will have to sign a breaching memo issued by the village administration and any individual benefits the community member's household would receive from the Village Development Fund will be reduced by 50% for 3 months.
- Any further transgression: The community member's household will be excluded for 6 months from any individual benefits from the Village Development Fund.
- Additionally, penalties stipulated in the Lao PDR legislation may apply.

Article 4: penalties for violations of relevant regulations and laws of Lao PDR

Seven violations with corresponding level of fines are stipulated and prosecution will be applied.

[Violations *and fines listed in Lao version.*]

Implementation of the penalties stipulated above will be the responsibility of relevant government agencies.

Article 5: village development fund

5.1 The Village Head and the Village Development Committee agree to manage the Village Development Fund according to the principles established by the Luxembourg Agency of Development Cooperation. The money in the Village Development Fund will be used in accordance with the Community Action Plan.

5.2 The Village Head and the Village Development Committee agree to support the training of the managers of the Village Development Fund.

Article 6: mechanism for grievance, conflict resolution, and redress

6.1 All villagers have the right to solve any grievance directly with the Patrol Manager. All villagers also have the right to file a complaint using the mechanism for grievance, conflict resolution, and redress without having first attempted to resolve the dispute directly with the Patrol Manager.

6.2 The Village Head and the Village Development Committee agree to establish a Village Mediation Unit. The Village Mediation Unit will consist of a selected group of villagers, including ethnic minorities, women, and representatives of other vulnerable groups in the village.

6.3 The Village Head and the Village Development Committee agree that the Village Mediation Unit will be in charge of managing the grievance process at the village level as outlined in the Community Action Plan.

6.4 The Village Head and the Village Development Committee agree to the following:
   a. All members of the Village Mediation Unit will participate in a training workshop that will enable them to perform the tasks involved with managing the mechanism for grievance, conflict resolution, and redress.
   b. The Village Mediation Unit will inform all villagers about their right to submit grievances to the mechanism of grievance, conflict resolution, and redress.
   c. The Village Mediation Unit will explain to all villagers how the mechanism of grievance, conflict resolution, and redress works.
   d. The Village Mediation Unit will address adequately concerns of all villagers related to the Village Development Fund, the Village Conservation Agreement or the antipoaching patrol scheme in a prompt and timely manner.

6.5 The Village Head and the Village Development Committee agree to offer access to the mechanism of grievance, conflict resolution, and redress to all community members free of charge.

Article 7: monitoring and evaluation

7.1 The Village Head and the Village Development Committee agree to hold quarterly meetings organized by the Village Development Committee. During these meetings, all villagers will have the opportunity to discuss concerns, suggest improvements, and report any negative impacts that have occurred but not yet resolved. The Village Development Committee will pay particular attention that vulnerable people and minority ethnic groups have the opportunity to participate and voice their opinions and concerns during the village level meetings. Representatives of the Village Mediation Unit will participate in these meetings to discuss outstanding grievances or issues that have been raised.

7.2 The Village Head and the Village Development Committee agree that two representatives of the village will attend an annual meeting at the district level. One representative will be a member of the Village Development Committee, the other a villager nominated by your community. If the Village Development Committee representative is a male, then the second village representative will be a female, or the vice versa. The representatives will be encouraged to share their community's perspectives on the performance of the PES scheme, suggestions for improvement, outstanding grievances, and other relevant issues. Measures to improve performance and resolve outstanding grievances will be also discussed and agreed on. The meeting organizers will take minutes. Progress on actions agreed at the meeting will be discussed in the meeting to be held in the following year.

This Community Agreement is effective as of the signed date onward. Signatories:

Village Head

Faculty of Economics and Business Management, National University of Laos
Certify by:
The District Governor of [district name]

Patrol contract template

Lao People's Democratic Republic

Peace Independence Democracy Unity Prosperity

Province: Bolikhamxay

District: [district name]

Village: [village name]

No.:   /

Date:

On the protection of biodiversity in the Phou Chomvoy Provincial Protected Area

- Pursuant to the Forestry Law No. 06/NA, dated December 24, 2007.
- Pursuant to the Law on Wildlife and Aquatic Animals No. 07/NA, dated December 24, 2007.
- Pursuant to the Decree on Protected Areas No. 134/Government, dated May 13, 2015.
- Pursuant to the Agreement of the Vice Prime Minister, Minister of Finance, Vice President of the Environment Protection Fund's Board of Management regarding the Approval of the EPF Subproject funding supported by the World Bank No. 008/BoM.EPF, dated July 8, 2016.
- Pursuant to the Agreement of the Governor of Bolikhamxay Province No. 516/ BXP dated July 14, 2017 on Permission for the Faculty of Economics and Business Management, National University of Laos to Conduct a Research Project on Implementation of Payments for Environmental Services in the Phou Chomvoy Protected Area in Bolikhamxay Province.

195

This Patrol Contract, is made and entered into by [*insert patrol team members' names*] (hereinafter "Patrol Team") and the Faculty of Economics and Business Management, National University of Laos.

The content of this Patrol Contract was developed with the agreement and acknowledgment of all relevant Lao PDR government authorities at the village, district, provincial, and national level. [*Insert village name*] has agreed to support the implementation of an antipoaching patrolling scheme under which the Patrol Team will operate. This agreement was formalized through a Community Conservation Agreement between the [*insert village name*] and the Faculty of Economics and Business Management, National University of Laos.

An oath ceremony will be conducted with the assistance of monks and/or spiritual leaders tailored to each ethnic culture represented in the village to support this Patrol Contract. The oath will contain the following words: "We, the patrol team (each says his own name) oblige to take the commitment of anti-poaching patrolling in PCPPA with high responsibility. We swear that we will honestly carry out our duties as outlined in the Patrol Contract and will strictly follow the Environmental Code of Conduct. We acknowledge that should we breach the Patrol Contract penalties will apply."

This Patrol Contract may be modified if any of its conditions are not suitable to the situation. All parties that have signed the present agreement will be informed of and will acknowledge and agree on all changes and modifications done.

In consideration of these premises, the Patrol Team and the Faculty of Economics and Business Management, National University of Laos, intending to be legally bound, do hereby agree as follows:

Article 1: Definition of terms

For the purpose of this Patrol Contract, the following terms have the following meanings:

a. "Poaching" means any illegal hunting or catching of wildlife within the Phou Chomvoy Provincial Protected Area.

b. "Patrol" means seven consecutive days of antipoaching patrolling within the Phou Chomvoy Provincial Protected Area including travel no more than 2 days from and to the Patrol Team residence. During each patrol, a preset series of geographically designated grid cells has to be visited. The series of grid cells and the patrol pattern will be assigned by the Patrol Manager and will change for every patrol. The required number of grid cells will vary by seasons (less during the wet season than during the dry season) and by differences in the difficulty of terrain (less in densely vegetated and steeper areas than on flatter and less densely vegetated areas).

c. "Patrol Manager" means the person employed by the Wildlife Conservation Association who manages the antipoaching patrol scheme at the technical level.

d. "Busy season" means the months of the rice planting and harvesting season (June, July, October, and November).
e. "Quiet season" means the remaining months (January, February, March, April, May, August, September, and December).
f. "Physical and Cultural Resources" mean movable or immovable objects, sites, structures, groups of structures, and natural features and landscapes that have archeological, paleontological, historical, architectural, religious, aesthetic, or other cultural significance.

Article 2: Patrol Contract period

The Patrol Contract period of 3 years will commence on [*insert date*]. The Patrol Contract can be terminated prematurely if both parties agree.

Article 3: Description of service

The Patrol Team consists of five people (of which two or three are village militia).The Patrol Team will conduct [*insert number*] patrols during the busy season and [*insert number*] patrols during the quiet season. The patrols within each season will be scheduled by the Patrol Manager in consultation with the Patrol Team.

The Patrol Manager will also be in charge of planning and managing the location, timing, and tasks of the Patrol Team. The Patrol Manager will ensure that the Patrol Team is properly prepared, receive their assignment and the appropriate permission to do the patrols.

The Patrol Team will receive professional training and an operation manual produced by the Patrol Manager that will enable them to undertake the patrol tasks.

The Patrol Team will undertake the following tasks while patrolling:

a. Dismantling of snare lines and collection of snare wires. The Patrol Team will take photographs of encountered snare lines and snare wires in situ and dismantled as with a photo stamp showing the GPS location and date and time.
b. The Patrol Team will carry the snare wires back to the village.
c. Dismantling of illegal poaching camps. The Patrol Team will take photographs of encountered camps in situ and dismantled as with a photo stamp showing the GPS location and date and time.
d. Recording poaching incidents and evidence. The Patrol Team will take notes and photographs of encountered poaching incidents and evidence with a photo stamp showing the GPS location and date and time of the encounter.
e. Reporting encountered poachers. The Patrol Team will record conversations with poachers, request signed statements of poachers, and alert the Provincial Office of Forestry Protection (through SPIRIT) and the Provincial Agricultural and Forestry Office.
f. Confiscating illegal gear (guns, rifles) used for poaching wildlife. In case of risk of violent clashes, on-spot confiscation of guns and rifles is not recommended

other than taking the evidence and reporting it to the police. The Patrol Team will take photographs of confiscated gear and the poacher(s) with a photo stamp showing the GPS location and date and time of the confiscation. The Patrol Team will ask poachers to sign a confiscation form, which the Patrol Team will also sign. The Patrol Team will hand over confiscated poaching gear to the police.

g. Confiscation of animals and animal parts, including dead animals/parts, and release of animals that are alive. In case of risk of violent clashes, on-spot confiscation of animals and animal parts is not recommended other than taking the evidence and reporting it to the police. The Patrol Team will take photographs of confiscated animals, animal parts, and the poacher(s) with a photo stamp showing the GPS location and date and time of confiscation. The Patrol Team will ask poachers to sign a confiscation form, which the Patrol Team will also sign.

h. Issuing of warnings to local poachers. The Patrol Team will issue an official warning using warning forms that need to be signed by the local poachers. The Patrol Team will record relevant information.

i. Apprehending poachers who are not Lao PDR citizen. In case of risk of violent clashes, apprehending poachers is not recommended other than taking the evidence and reporting it to the police. The Patrol Team will turnover apprehended poachers to the police and alert the Provincial Office of Forestry Inspection (through SPIRIT) as well as the Provincial Agricultural and Forestry Office. Any expenses to village authorities are beyond the responsibility of the PES scheme.

j. Recording of any direct sightings of key wildlife species as specified by the Patrol Manager. The Patrol Team will record information on the species, number of individual animals, as well as the GPS location and date and time of sightings using a data form.

k. Record of any indirect signs (e.g., tracks, scat) of key wildlife species every 300 m. The Patrol Team will record information on the species as well as the GPS location and date and time of signs using a data form.

In any circumstance where a violent confrontation is likely, the Patrol Team should withdraw from immediate engagement.

The Patrol Team will comply with the Environmental Code of Conduct. The Environmental Code of Conduct will ensure that patrolling in the PCPPA will damage neither wildlife nor their habitat. The Environmental Code of Conduct consists of the following rules:

a. The Patrol Team is not allowed to hunt any wildlife for food during antipoaching patrolling.

b. The Patrol Team will minimize disturbance of wildlife through, for example, loud noise.

c. The Patrol Team will not make camp in ecologically fragile areas and will minimize the cutting of vegetation and site clearing.

d. The Patrol Team will dismantle their camps and putting off their cooking fire completely before continuing the patrol.
e. Cigarette stubs must be extinguished completely.
f. The Patrol Team will carry out any garbage (including cigarette stubs) and not discard it within the PCPPA.
g. The Patrol Team is expected to be a role model for the wider community. Breeching the Community Conservation Agreement by any member of the Patrol Team during antipoaching patrolling or off-duty will be treated as a violation of this Environmental Code of Conduct.
h. The Patrol Team is not allowed to drink alcohol during patrol activities.

The Patrol Team will comply with the Physical and Cultural Resources Chance-Find Procedure. Chance-Find Procedures have been developed to mitigate against damage or loss to Physical and Cultural Resources. Chance-Find Procedures relevant to the Patrol Team include the following:

a. The Patrol Team will not move or interfere with a suspected Chance-Find.
b. The Patrol Team will immediately report a suspected Chance-Find to the Village Head and a representative of the District Agricultural and Forestry Office.

Article 4: Terms of payment

a. Price per Patrol. The Patrol Team will receive LAK 60,000/day/person per completed Patrol. A Patrol is recognized as completed if the Patrol Team provides full and valid evidence that the assigned series of grid cells was visited and the assigned tasks were fulfilled. Evidence that needs to be presented to the Patrol Manager consists of recorded GPS coordinates as well as of photographs of team members a photo stamp showing the GPS location and date and time of the patrol team at the start, middle, and end of each patrol day. Patrol teams get paid in full if they complete their tasks within the 7-day patrol period. A deviation by plus or minus 1 day is acceptable.
b. Price per snare wire. The Patrol Team will receive, conditional on full and valid evidence, LAK 2000 for each used snare wire collected during patrolling. Evidence that needs to be presented to the Patrol Manager consists of photographs of the snare line before the dismantling and after as well as of the collected snare wires with a photo stamp showing the GPS location and date and time. As the density of the snares decreases, the price per snare wire will be adjusted upwards at the discretion of the Patrol Manager.
c. Price per dismantled camp. The Patrol Team will receive, conditional on full and valid evidence, LAK 10,000 for each camp dismantled during patrolling. Evidence that needs to be presented to the Patrol Manager consists of photographs of the camp before the dismantling and after with a photo stamp showing the GPS location and date and time. As the density of camps decreases, the price per camp will be adjusted upwards at the discretion of the Patrol Manager.

d. Payment schedule and transfer. An advance of 30% on the first regular payment will be paid to the Patrol Team immediately after signing this Patrol Contract. Subsequent payments to the Patrol Team will be made every 3 months. The Patrol Manager will calculate the payments every 3 months on the performance of the Patrol Team and will then send a payment request to the Faculty of Economics and Business Management at the National University of Laos. The Patrol Team is encouraged to calculate their own payments to cross-check the calculations made by the Patrol Manager. The Faculty of Economics and Business Management at the National University of Laos will approve the requested amount of payments by withdrawal from its PES scheme bank account in Vientiane. The Patrol Manager will bring the cash to pay the patrol teams. Receipt of each payment will be signed by each patrol member and certified by the village head.

Article 5: Equipment

a. Provision. The Patrol Team will receive the following list of equipment required to fulfill the Patrol commitments. The first set of equipment provided will remain the property of the National University of Laos: one camera with in-built GPS function (if required), one GPS unit (if required), one SIM-card for mobile phone or radio (if required), maps, one compass, one pair of binoculars, one paper notebook and record sheets, five backpacks, five flysheets, five hammocks, and one first aid kit. The second set of equipment provided will become the property of the Patrol Team: mosquito repellent, five pairs of antileech socks, five hats, field clothing (pants and shirt marked as "village patrol") for five people, and five pairs of boots. The management of the equipment will be specified in the operation manual issued to the Patrol Team by the Patrol Manager.

b. Malfunction. Malfunctioning equipment will be repaired or replaced if necessary by the Patrol Manager who will request funding from the Environmental Protection Fund.

c. Loss. A sliding penalty for each equipment category will be applied to the Patrol Team for the replacement of lost or damaged (caused by users beyond expected wear and tear) items of equipment that is the property of National University of Laos. After the first loss, without a valid reason or due to user's fault, 25% of replacement costs will be paid by the Patrol Team (taken out of the Patrol Team's payments). After the second loss, 50% of replacement costs will be paid by the Patrol Team (taken out of the Patrol Team's payments). After the third and any further loss, 100% of replacement costs will be paid by the Patrol Team (taken out of the Patrol Team's payments). Lost equipment that became property of the Patrol Team will be replaced by the Patrol Team members.

Article 6: Recognition

Each Patrol Team member will be recognized as a "Trusted Biodiversity Guardian" if his or her team has fulfilled all its commitments and made exemplarily efforts over the course of 1 year. Each Patrol Team member will receive a certificate and a badge for his or her field clothing presented to him or her by a high-ranking government official.

Article 7: Other benefits

The Patrol Team may also be eligible for rewards form the Government of Lao PDR for activities that support wildlife law enforcement This Patrol Contract does not preclude the Patrol Teams from negotiating such rewards with the relevant government authorities.

Article 8: Insurance cover

The Patrol Team will receive health and accident insurance that covers their patrolling activities.

Insurance for medical expenses following illness and minor accidents will be provided through a commercial insurance company arranged through the Environmental Protection Fund. The insurance will cover benefits as specified in the insurance policy of the insurance company.

A one-off compensation payment in case of death and permanent disability as well as medical expenses, including transport to a formal hospital within the country, following a major accident or a violence related incident are covered through the contingency budget earmarked for accident insurance managed by Environmental Protection Fund. The Environmental Protection Fund will cover the following benefits:

a. Medical expenses (up to LAK 50,000,000) for transport to and required treatment in a formal clinic or hospital inside the country following a major accident or violent incident that occurred while conducting antipoaching patrolling.
b. One-off compensation payment (LAK 50,000,000) in case of permanent disability following an accident or violent incident that occurred while conducting antipoaching patrolling.
c. One-off compensation payment (LAK 50,000,000) in case of death following a major accident or violence incidence that occurred while conducting antipoaching patrolling.

Any illness or injury that requires hospital care has to be reported to the Patrol Manager who will then arrange with the Faculty of Economics and Business Management at the National University of Laos to cover the medical costs.

Traditional healing and treatments outside a formal clinic or hospital are not covered.

Article 9: Penalties for breaching contractual obligations

The Patrol Team will only be paid for completed Patrols. If the Patrol Team does not provide the full set of evidence relating to its patrol activity, it will not be paid for the Patrol.

If the Patrol Team does not complete the number of Patrols scheduled by the Patrol Manager within a 3-month period, the three-monthly payment (calculated as the number of completed Patrols times the price per Patrol) will be reduced by 20%.

The payment per Patrol is the sum of every Patrol Team members' portion. Patrol Team members who do not complete the Patrol will not be paid their portion. That is, the Patrol Team payment will be reduced accordingly.

In case a Patrol Team member gets sick during patrolling and needs to be walked out by a fellow Patrol Team member, the Patrol Team will be paid in full for that Patrol if the following condition is met: The sick Patrol Team member can provide proof of illness (consisting of a testimonial from the Team Leader and a certificate from the Village Head or a health post/hospital) and if the rest of the team fulfills the missing Patrol Team members' tasks.

If valid proof is not presented and/or the remaining Patrol Team members do not fulfill the missing Patrol Team members' tasks, the team payment will be reduced to the portions of the remaining number of Patrol Team members who patrolled and the days they patrolled.

If the Patrol Team does not provide the full set of evidence for dismantling a camp, it will not get paid for the dismantling of the camp.

If the Patrol Team does not provide the full set of evidence for dismantling a snare line, it will not get paid for the collected snares.

If a Patrol Team member breaches the Environmental Code of Conduct, the following penalties will apply:

a. First transgression: The Patrol Team member will be given a verbal warning by the Village Head in the presence of the Patrol Manager.
b. Second transgression: The Patrol Team member will have to sign a breaching memo issued by the village administration and the Patrol Team's next payment will be reduced by 50% of the Patrol Team member's portion. The Patrol Team member's portion will be calculated as the number of Patrols the Patrol Team member performed within the current 3-month payment period multiplied by a fifths of the price per Patrol.
c. Third transgression: The Patrol Team's next payment will be reduced by 100% of the Patrol Team member's portion. The Patrol Team member's portion will be calculated as the number of Patrols the Patrol Team member performed within the current 3-month payment period multiplied by a fifths of the price per Patrol.
d. Any further transgression: The Patrol Team member will be excluded from the Patrol Team for the rest of the contract duration.

Additionally, penalties stipulated by the Lao legislation may apply.

Article 10: Mechanism for grievance, conflict resolution, and redress

The Patrol Team or any Patrol Team member has the right to solve any grievance directly with the Patrol Manager. The Patrol Team or any Patrol Team member also have the right to file a complaint using the mechanism for grievance, conflict resolution, and redress without having first attempted to resolve the dispute directly with the Patrol Manager. The use of this mechanism is free of charge and consists of the following steps:

a. Report the grievance to the Village Mediation Unit. The Village Mediation Unit will be in charge of documenting the grievance.
b. All grievances at the village level will be recorded with details, discussions, actions, and outcomes in a grievance logbook kept by the Village Mediation Units. The logbook will be accessible at all times to everyone involved.
c. If the grievance has not been resolved, the issue will be transferred to the next level, led by the District Steering Committee, for further action.
d. The District Steering Committee will also be in charge of compiling all grievances into a District Grievance logbook.
e. If the grievance has not been resolved, the issue will be transferred to the next level, led by the Provincial Steering Committee, for further action.
f. The Provincial Steering Committee will also be in charge of compiling all grievances into a Provincial Grievance logbook.
g. If the grievance has not been resolved, the issue will be transferred to the next level, led by the National Steering Committee, for further action.
h. Grievances can also be reported through the participatory Monitoring and Evaluation process or to the annual technical audit team as defined in the Community Conservation Agreement between the village and the Faculty of Economics and Business Management, National University of Laos.
i. Additionally, grievances can be reported directly to the National Steering Committee or the National Assembly.

Article 11: Prohibition

The Patrol Team is not allowed to subcontract any of the contracted services as set out in Article 3 (Description of service).

Article 12: Force majeure

Neither party will be liable for failure to perform, nor be deemed to be default, under this Patrol Contract for any delay or failure in performance resulting from causes beyond its reasonable control, including but not limited to failure of performance by the other party, acts of state or government authorities, acts of terrorism, natural catastrophe, fire, storm, flood, earthquake, riot, insurrection, civil disturbance, sabotage, embargo, blockade, acts of war, or power failure. In the event of such delay, the time of completion will be extended by a period of time reasonably necessary to overcome the effect of any such delay.

Article 13: Limitation of liability against a third party

In no event will either party be liable for special, indirect, consequential, or incidental damages to a third party, including but not limited to loss of profits, revenues, power, damage to or loss of the use of products, damage to property, claims of third parties, including personal injury and death, suffered as a result of provision of the contracted services set out in Article 3 (Description of service).

Article 14: General terms

This Patrol Contract is deemed to have been made, executed, and delivered in the Lao PDR and is constructed in accordance with the laws of the Lao PDR.

The invalidity or enforceability, in whole or in part, of any provision in this Patrol Contract will not affect in any way the remainder of the provisions herein.

This contract created by the two Parties with unanimous agreement is accepted and has an equal effect under the law and the Constitution of the Lao PDR. Should any provisions of this Contract contradict legislation and other agreements, the Contractual Parties shall together amend them to conform to the legal provisions. This agreement constitutes the final and entire Patrol Contract between the Patrol Team and the District Governor and supersedes all prior and contemporary agreements regarding antipoaching patrolling in the Phou Chomvoy Provincial Protected Area, oral or written which may contradict this agreement.

This Patrol Contract is effective as of the date stated on the first page. This Contract is signed in two duplicates for each Party to keep as evidence.

Signatories:

Patrol Team Members
Faculty of Economics and Business
Management
National University of Laos

Certify by
The Village Head of [insert village name]

Index

'Note: Page numbers followed by "t" indicate tables, "f" indicate figures and "b" indicate boxes.'

Printed in the United States
By Bookmasters